ENGINEERING CHEMISTRY

ENGINEERING CHEMISTRY

[Strictly according to Latest Syllabus of B.Tech. Ist Year, Jawaharlal Nehru Technological University, Kakinada]

By

Dr B.N. Srinivas
Ph.D. in Catalysis & Green Chemistry
Prof. and Head of the Department
of Chemistry
URCET, Telaprolu, Vijayawada
Andhra Pradesh

P. Kishore
Assistant Professor in
Department of Chemistry
URCET, Telaprolu, Vijayawada
Andhra Pradesh

K. Subba Rao
Assistant Professor in
Department of Chemistry
URCET, Telaprolu, Vijayawada
Andhra Pradesh

UNIVERSITY SCIENCE PRESS

(An Imprint of Laxmi Publications Pvt. Ltd.)

BANGALORE • CHENNAI • COCHIN • GUWAHATI • HYDERABAD
JALANDHAR • KOLKATA • LUCKNOW • MUMBAI • RANCHI • NEW DELHI
INDIA • USA • GHANA • KENYA

ENGINEERING CHEMISTRY

Printed and bound in India
Typeset at Goswami Associates, Delhi
First Edition: 2015
UEC-9742-150-ENGG CHEMISTRY-SRI
ISBN 978-93-83828-45-6

PUBLISHED IN INDIA BY

UNIVERSITY SCIENCE PRESS
(An Imprint of Laxmi Publications Pvt.Ltd.)
113, GOLDEN HOUSE, DARYAGANJ,
NEW DELHI - 110002, INDIA
Telephone : 91-11-4353 2500, 4353 2501
Fax : 91-11-2325 2572, 4353 2528
www.laxmipublications.com info@laxmipublications.com

Branches

©	Bangalore	080-26 75 69 30	
©	Chennai	044-24 34 47 26,	24 35 95 07
©	Cochin	0484-237 70 04,	405 13 03
©	Guwahati	0361-254 36 69,	251 38 81
©	Hyderabad	040-27 55 53 83,	27 55 53 93
©	Jalandhar	0181-222 12 72	
©	Kolkata	033-22 27 43 84	
©	Lucknow	0522-220 99 16	
©	Mumbai	022-24 91 54 15,	24 92 78 69
©	Ranchi	0651-220 44 64	

C—
Printed at : Repro Knowledgecast Limited, Thane

CONTENTS

PREFACE

Engineering Chemistry has been prepared exactly as per the latest syllabus prescribed by Jawaharlal Nehru Technological University, Kakinada, Andhra Pradesh. In this book, various topics are precisely explained, giving examples and figures wherever necessary. This book is aimed at imparting the required knowledge for understanding the importance of chemistry in the field of engineering.

The fundamental aspects of Osmosis, Electrochemistry, Corrosion, Polymers, Fuels were dealt with in a comprehensive way. Emphasis was given to the advanced topics like Nano-materials, Carbon nanotubes (CNTs), Fullerenes, Liquid Crystals, Fiber Reinforced Plastics (FRP), Carbon Fiber Reinforced Plastics (CFRP), Glass Fiber Reinforced Plastics, Bullet proof plastics, Biodegradable polymers, Conducting polymers, Solar cells, Solar panel, Green House Concept, Green chemistry, Cement comparative studies were also presented in the form of tables showing the differences between certain similar types of aspects.

We have done our job with great care and caution. However, there may be some printing errors, we request the teachers of Engineering Chemistry, to kindly inform us to correct them. Suggestions for further improvement will be highly appreciated and thankfully acknowledge.

—Authors

ACKNOWLEDGEMENTS

It is indeed our privilege and duty to thank Sri Sunkara Rama Brahmam, Chairman, Sri Sunkara Anil Kumar and Sri Sunkara Ajay, Secretary and Correspondents of **Usha Rama College of Engineering and Technology (URCET), Telaprolu (Krishna District),** without their encouragement, this might not have seen the light of the day.

We feel it is great pleasure in expressing our gratitude to the Principal, Dr G.V.K.S.V. Prasad, **Usha Rama College of Engineering and Technology, Telaprolu (Krishna District),** for his never ending inspiration and encouragement in our effort.

We feel happy to acknowledge the valuable suggestions and support extended by Dr T.S.R. Prasad Rao, **Retd. Director,** Indian Institute of Petroleum, Dehradun, Dr K.S. Rama Rao, Dr A. Venu Gopal, Dr N. Lingaiah, Dr P. Laxmi Kantham, Director, **IICT, Hyderabad.**

We sincerely thanks to our colleagues and friends for their valuable suggestions and moral support. We are very much obliged to our family members for their constant support, without which it would have been impossible to complete this task. We extend our sincere thanks for other references and their authors.

Finally we extend our gratitude to Laxmi Publications and their staff for their efforts without which the book could not have taken this shape today.

—Authors

SYLLABUS

ENGINEERING CHEMISTRY

UNIT 1: WATER TECHNOLOGY

Hard Water – Estimation of hardness by EDTA method – Potable water – Sterilization and Disinfection – Boiler feed water – Boiler troubles – Priming and foaming , scale formation, corrosion, caustic embrittlement, turbine deposits – Softening of water – Lime soda, Zeolite processes – Reverse osmosis – Electrodialysis, Ion exchange process.

Objectives: For prospective engineers knowledge about water used in industries (boilers etc.) and for drinking purposes is useful; hence chemistry of hard water, boiler troubles and modern methods of softening hard water is introduced.

UNIT 2: ELECTROCHEMISTRY

Concept of Ionic Mobilities – Applications of Kohlrausch law – Conductometric titrations – Galvanic cells – Electrode potentials – Nernst equation – Electrochemical series – Potentiometric titrations – Concentration cells – Ion selective electrode: Batteries and Fuel cells.

Objectives: Knowledge of galvanic cells, electrode potentials, concentration cells is necessary for engineers to understand corrosion problem and its control ; also this knowledge helps in understanding modern biosensors, fuel cells and improve them.

UNIT 3: CORROSION

Causes and effects of corrosion – theories of corrosion (dry, chemical and electrochemical corrosion) – Factors effecting corrosion – Corrosion control methods – Cathode protection –Sacrificial Anodic, Impressed current methods – Surface coatings – Methods of application on metals (Hot dipping, Galvanizing, tinning , Cladding, Electroplating, Electroless plating) – Organic surface coatings – Paints – Their constituents and their functions.

Objectives: The problems associated with corrosion are well known and the engineers must be aware of these problems and also how to counter them.

UNIT 4: HIGH POLYMERS

Types of Polymerization – Stereopolymers – Physical and Mechanical properties of polymers – Plastics – Thermoplastics and thermosetting plastics – Compounding and Fabrication of plastics – Preparation and properties of Polyethylene, PVC and Bakelite – Elastomers – Rubber and Vulcanization – Synthetic rubbers – Styrene butadiene rubber – Thiokol – applications.

Objectives: Plastics are materials used very widely an engineering materials. An understanding of properties particularly physical and mechanical properties of polymers/plastics/ elastomers helps in selecting suitable materials for different purpose.

UNIT 5: FUELS

Coal – Proximate and ultimate analysis – Numerical problems based on analysis – Calorific value – HCV and LCV – Problems based on calorific values; petroleum – Refining – Cracking – Petrol – Diesel knocking; Gaseous fuels – Natural gas – LPG, CNG – Combustion – Problems on air requirements.

Objectives: A board understanding of the more important fuels employed on a large scale is necessary for all engineer to understand energy – related problems and solve them.

UNIT 6: CHEMISTRY OF ADVENCED MATERIALS

Nanomaterials (Preparation of carbon nanotubes and fullerenes – Properties of nanomaterials – Engineering applications) – Liquid crystals (Types – Application in LCD and Engineering Applications) – Fiber reinforced plastics – Biodegradable polymers – Conducting polymers – Solar cells (Solar heaters – Photovoltaic cells – Solar reflectors – Greenhouse concepts – Green chemistry (Methods for green synthesis and Applications) – Cement – Hardening and setting – Deterioration of cement concrete.

Objectives: With the knowledge available now, future engineers should know at least some of the advanced materials that are becoming available. Hence some of them are introduced here.

Chapter 1

WATER TECHNOLOGY

1.1 INTRODUCTION

Water is the most useful, wonderful and abundant compound on earth. It is a vital component of the living sources. It is a proven theory that one can live without food for many number of days but one cannot live without water. Water is the important constituent of all the body fluids, without which all the cells or organisms are in crystalline or dead state. (For example, the human body contains 70%, terestrial plants 50 – 75% and aquatic plants 95 – 99%). Water is not only essential for the lives of animals and plants, but also occupies a unique position in industry.

Industrial Purpose of Water:

- Water is used in power generation, as steam in steam turbines and as coolants.
- Blasting and water jet cutters. Very high pressure water guns are used for precise cutting. It is also used in the cooling of machinery to prevent over-heating, or prevent saw blades from over-heating.
- Industry requires pure water for many applications and utilizes a variety of purification techniques both in water supply and discharge.
- Water plays many critical roles in the field of food science.
- Water is widely used in the production of steel, rayon, paper, atomic energy, textiles, chemicals, ice and for air conditioning, drinking, bathing, sanitary, washing, irrigation and fire-fighting etc.

- **Sources of water.** The chief sources of water supply for industrial use are:

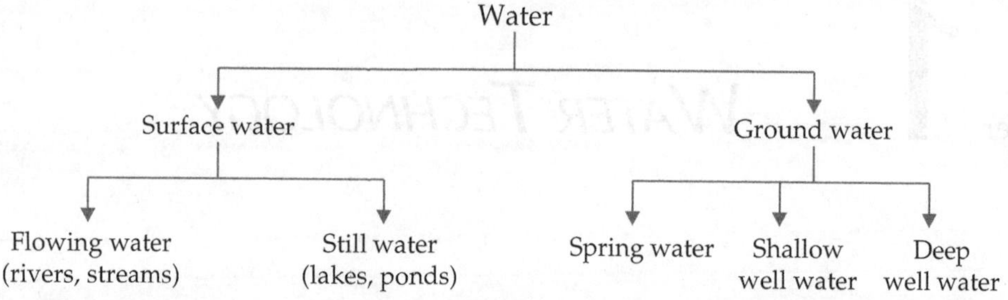

FIGURE 1.1

- **Rain water.** The purest form of water, collected on the roofs. Yet this method is seldom adopted in industry.
- **Surface waters.**
  - ➢ Flowing water, such as rivers, streams etc.
  - ➢ Still water, such as lakes, ponds etc.
- **Ground water.**
  - ➢ Water from springs.
  - ➢ Water from shallow wells. In the case of shallow wells, the boring is done only through one geological stratum.
  - ➢ Water from deep wells. Here the boring is done through many geological strata.
- **Sea water.** Its use is very limited as its use many problems of chemical engineering.

Source and Nature of Impurities of Water. Water in the form of vapour in clouds is said to be pure. Yet when it condenses as rain and flows on the ground it takes many impurities from atmosphere and ground. The major situations in which water gets impure are as follows:

- **Dissolved gases.** During raining water absorbs much of the gases like oxygen, carbon dioxide, hydrogen sulphide etc. from atmosphere. The resultant water will be slightly acidic and on high concentrations of impurities it may result into acid rain.
- **Soluble crystalloids.** Water when flows over the surface of the land (like rivers, streams) dissolves soluble minerals. The most of the soluble minerals include chlorides, sulphates and bicarbonates of sodium, calcium, magnesium and iron.
  - ➢ Whereas in seawater 3.5% of dissolved salts are seen in which 2.5% include sodium chloride.
- **Organic matter.** This is derived from the decomposition of plants and small particles of sand and rock in suspension.

Characteristics Imparted by Impurities of Water. As such pure water is odourless, colourless, tasteless, etc., but the presence of impurities imparts various physical and chemical changes to water nature which are as follows.

1. Colour. The metallic substances like humus, tannins, peat, algae, weeds, protozoa, industrial effluents and salts of iron, manganese in the colloidal form imparts colour to water.

For example, yellowish tinge is due to the presence of chromium or appreciable amount of organic matter, yellowish–red colour due to the presence of iron, while reddish brown is due to the presence of peat. (The intensity of the colour indicates the severity of toxic nature of water).

2. Turbidity. The colloidal, extremely fine suspension particles such as clay, silt, finely divided matter, micro-organisms etc. increases the turbidity of the water. The turbid water is not fit for either consumption or industrial purpose.

3. Taste. Usually this property is linked to odour. Yet under certain conditions taste is not accompanied by odour. For example,

- Bitter taste can be due to the presence of iron, aluminum, manganese, sulphate or excess of lime.
- Soapy taste can be due to the presence of large amount of sodium bicarbonate.
- Brackish taste is due to the presence of unusual amount of salts.
- Palatable taste is due to the presence of dissolved gases (CO_2) and minerals like nitrates.

4. Odour. The odour may be due to the presence of living organisms, decaying vegetation. The common disagreeable odour is due to the presence of sulphides. The odour of sewage water is due to the presence of organic compounds of N, S and P along with the putrefaction of proteins and the industrial effluents like alcohols, aldehydes, phenols, esters, ketones etc. Besides these,

- The presence of algae in water impart strong grassy odour.
- Growth of iron and sulphur bacteria produces offensive odour.
- Sand in finely divided state imparts faint earthy odour.
- When water passes through iron pipes, it gives tarry odour.
- When the water passes through galvanized iron pipes impure hydrogen will be liberated which results infaint odour.

5. Chemical impurities in water. Chemical impurities in water includes:

- Inorganic and organic chemicals released from dyes, paints, varnishes, drugs, insecticides, pesticides, detergents and textiles, industries, etc. All these pollute the water bodies.
- Acid discharged in water by DDT, high explosives, battery, industries, etc. The use of this type of contaminated water causes harmful effects on health of human-beings.

6. Biological impurities. Biological impurities are algae, pathogenic bacteria, fungi, viruses, pathogens, etc. The source of contamination is discharge of domestic and sewage wastes, excreta (from man, animals and birds), etc.

1.2 HARDNESS OF WATER

Hardness in water is the character, which "prevents the lathering of soap". This is due to the presence of bicarbonates, chlorides and sulphates of calcium, magnesium and other heavy metals like strontium and iron, etc.

For example, insoluble carbonates of Ca, Mg and Fe readily transform into soluble bicarbonates in the presence of carbon dioxide.

$$\underset{\text{Insoluble}}{CaCO_3} + CO_2 + H_2O \longrightarrow \underset{\text{Soluble}}{Ca(HCO_3)_2}$$

$$\underset{\text{Insoluble}}{MgCO_3} + CO_2 + H_2O \longrightarrow \underset{\text{Soluble}}{Mg(HCO_3)_2}$$

Many of the mineral constituents of rocks like NaCl, $CaSO_4 . H_2O$ readily dissolve in water and collects in it.

$$\underset{\text{Anhydrite}}{CaSO_4} + 2H_2O \longrightarrow \underset{\text{Gypsum}}{CaSO_4 . 2H_2O}$$

If a sample of water is treated with soap (Na/K salt of higher fatty acids like oleic, palmitic/stearic) like sodium streates and if the water does not produces lather, then the water is said to be hard water, else soft water.

In the hard water the soap produces precipitate or white scum, which is due to the formation of insoluble soap of calcium and magnesium. For example, a typical soap (Sodium Streate) reaction with salts like calcium chloride and magnesium sulphate are depicted as follows.

$$\underset{\text{Sodium streate}}{2C_{17}H_{35}COONa} + \underset{\text{Hardness}}{CaCl_2} \longrightarrow \underset{\text{Insoluble PPT}}{(C_{17}H_{35}COO)_2\,Ca\downarrow} + 2NaCl$$

$$\underset{\text{Sodium streate}}{2C_{17}H_{35}COONa} + \underset{\text{Hardness}}{MgSO_4} \longrightarrow \underset{\text{Insoluble PPT}}{(C_{17}H_{35}COO)_2\,Mg\downarrow} + Na_2SO_4$$

1.3 TYPES OF HARDNESS

Hardness of water is of two types:

1. Temporary or Carbonate hardness
2. Permanent or Non-carbonate hardness.

1. Temporary or Carbonate Hardness. Temporary hardness of water is due to the presence of dissolved bicarbonates of calcium, magnesium and other heavy metals or carbonate of iron. It is easily destroyed by mere boiling of water. When the water is boiled, the bicarbonates are decomposed, yielding insoluble carbonates or hydroxides which are deposited as a crust or scales at the bottom of the vessel, while carbon dioxide escapes out.

$$\underset{\text{Soluble}}{Ca(HCO_3)_2} \longrightarrow \underset{\text{Insoluble}}{CaCO_3\downarrow} + H_2O + CO_2\uparrow$$

$$\underset{\text{Soluble}}{Mg(HCO_3)_2} \longrightarrow \underset{\text{Insoluble}}{Mg(OH)_2\downarrow} + 2CO_2\uparrow$$

2. Permanent or Non-carbonate Hardness. Permanent hardness is due to the presence of chlorides, sulphates of calcium, magnesium, iron and other heavy metals. It cannot be removed by boiling; it is removed by using softeners or chemical treatment.

Disadvantages of hard water:

1. Domestic:

- **Washing and bathing.** Hard water does not form lather easily with soap. As a result, a large amount of soap is wasted.

- **Drinking.** Hard water causes bad effects on our digestive system. Sometimes, stone formation takes place in kidneys.
- **Cooking.** The boiling point of water is increased due to the presence of salts. Hence, more fuel and time are required for cooking.

2. Industrial:

- **Textile industry.** Hard water causes wastage of soap. Precipitates of calcium and magnesium soaps adhere to the fabrics and cause problem.
- **Paper industry.** Calcium and magnesium salts in water may affect the quality of paper.
- **Sugar industry.** Water containing sulphates, carbonates, nitrates affects the crystallization of sugar.
- **Dyeing industry.** The salts of calcium and magnesium in hard water react with dyes and spoil the desired shade.
- **Pharmaceutical industry.** Hard water may form some undesirable products while preparation of pharmaceutical products.
- **Concrete making.** Chloride and sulphates present in hard water will affect the hydration of cement and the final strength of the hardened concrete.
- Finally in industries where steam is employed, if hard water is used in steam production, the troubles like corrosion, scale and sludge formation, priming and foaming and caustic embrittlement are seen.

Expression of Concentration of Hardness. The concentration of hardness as well as non-hardness constituting ions is expressed in terms of an equivalent amount of $CaCO_3$. The equivalence of $CaCO_3$ is used because it permits the addition and subtraction of concentration, when required. The choice of $CaCO_3$ in particular is due to its molecular weight 100 (equivalent wt. 50). Not only that, it is the most insoluble salt that can be precipitated in the water treatment easily.

The equivalents of $CaCO_3$ can be defined as:

The equivalents of $CaCO_3$

$$= \frac{[\text{Mass of hardness producing substance}] \times [\text{Chemical equivalent of } CaCO_3]}{\text{Chemical equivalent of hardness producing substance}}$$

$$= \frac{\text{Mass of hardness producing substance} \times 50}{\text{Chemical equivalent of hardness producing substance}} \text{ in mg/L}$$

1.4 UNITS OF EXPRESSION OF HARDNESS OF WATER

Normally the hardness of water is represented in ppm. *i.e.*, parts per million or mg/L.

$$1 \text{ mg/L} = 1 \text{ ppm.}$$

1. **Parts per million (ppm).** It is the number of parts by weight of $CaCO_3$equivalent hardness present per million (10^6) parts by water

   i.e., 1 ppm = 1 part of $CaCO_3$ equivalent hardness in 10^6 parts of water.

2. **Milligram/Litre (mg/L).** It is the number of milligrams of calcium carbonate equivalent present per litre of water.

i.e., 1 mg/L = 1 mg of $CaCO_3$ equivalent hardness per litre of water.

But, it is known that 1 lit = 1 kg = 1000 g = 1000 × 1000 mg = 10^6 mg.

⇒ 1 mg/L = 1 mg of $CaCO_3$ equivalent hardness per 10^6 mg of water
= 1 part of $CaCO_3$ equivalent hardness per 10^6 mg of water
= 1 ppm.

Hardness is also expressed in the following ways:

3. **Degree Clark (°Cl).**

°Cl = 1 part of $CaCO_3$ equivalent hardness per 70,000 parts of water

i.e., 1 ppm = 0.07°Cl

4. **Degree French (°Fr).**

°Fr = 1 part of $CaCO_3$ equivalent hardness per 10^5 parts of water

i.e., 1 ppm = 0.1°Fr

1.5 DETERMINATION OF HARDNESS OF WATER

The degree of hardness of water is estimated by the following ways:

1. Soap titration method.
2. O. Hener's alkalimetric method.
3. EDTA method.

1.5.1 Soap Titration Method

It is a simple technique. Soft water gives lather with soap whereas hard water does not. The dissolved salts in the hard water react with soap and are precipitated. Thus the amount of soap initially consumed before the formation of lather is a measure of the total hardness and indirectly the dissolved magnesium and calcium salts. The hardness is easily estimated by adding a soap solution of a definite strength to a known volume of the sample of water until a permanent lather is formed on shaking. The standard solution of soap can be obtained from the market or otherwise it is prepared and then standardized in the laboratory with standard $CaCl_2$ solution

$$2C_{17}H_{35}COONa + Ca^{++} \longrightarrow (C_{17}H_{35}COO)_2\,Ca\downarrow + 2Na^+$$
$$2C_{17}H_{35}COONa + Mg^{++} \longrightarrow (C_{17}H_{35}COO)_2\,Mg\downarrow + 2Na^+$$

1.5.2 O. Hener's Alkalimetric Method

Temporary hardness of water can be determined by using HCl or H_2SO_4, as the calcium or magnesium bicarbonates react directly with dilute acids.

$$Ca(HCO_3)_2 + 2HCl \longrightarrow CaCl_2 + 2H_2O + 2CO_2$$
$$Mg(HCO_3)_2 + 2HCl \longrightarrow MgCl_2 + 2H_2O + 2CO_2$$

The water sample is titrated with dilute hydrochloric acid solution taking in a burette after adding methyl orange indicator. The change of colour from yellow to pale pink indicates the end point. From the titre value the amount of temporary hardness in terms of $CaCO_3$ equivalents is calculated.

1.5.3 Estimation of Hardness in Water by EDTA Method

EDTA is Ethylene diamine tetra acetic acid whose structural formulae is

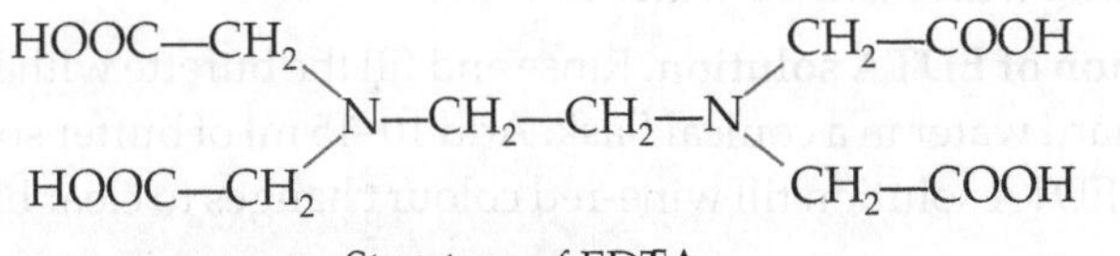

Structure of EDTA

In the form of its sodium salt yields the anion;

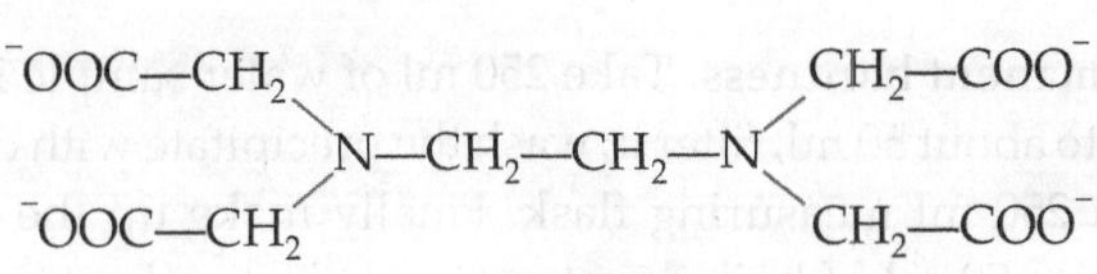

EDTA ion

Which forms complex ions with Ca^{+2} and Mg^{+2}?

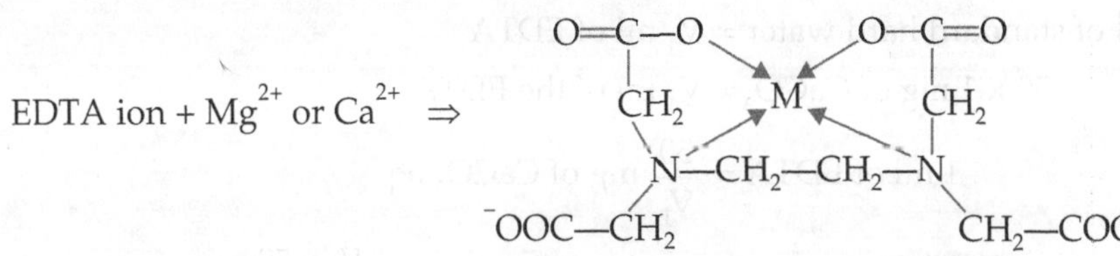

where M = Ca or Mg. It may be pointed that the EDTA is used as its disodium salt.

Principle. In this complex metric Titration, the water sample is titrated with standard solution of di sodium salt of EDTA using EBT indicator.

Reactions involved. In order to determine the equivalence point, indicator Eriochrome black-T (EBT) is used. When EBT is added to hard water at a pH of about 10, using ammoniacal buffer solution ($NH_4OH—NH_4Cl$),a wine red unstable complex with Ca^{2+}or Mg^{2+} ions is formed. When this solution is titrated with EDTA, at the endpoint wine red colour changes to deep blue colour solution *i.e.*, the EDTA has formed stable complexes.

$$M^{2-} + EBT \longrightarrow \underset{\text{Unstable wine red}}{M\text{-}EBT} + EDTA \longrightarrow \underset{\text{Stable deep blue}}{M\text{-}EDTA + EBT}$$

Thus, change of wine-red colour to distinct blue marks the end-point of titration.

Various steps involved in this method are:

1. Preparation of standard hard water. Dissolve 1 g of pure, dry $CaCO_3$ in minimum quantity of dil. HCl and then evaporate the solution to dryness on a water bath. Dissolve the residue in distilled water to make 1 litre solution. Each ml of this solution thus contains 1 mg of $CaCO_3$ equivalent hardness.

1 ml hard water solution = 1 mg of $CaCO_3$ equivalent hardness.

2. Preparation of EDTA solution. Dissolve 4 g of pure EDTA crystals + 0.1 g $MgCl_2$ in 1 litre of distilled water.

3. Preparation of indicator (EBT). Dissolve 0.5 g of Eriochrome Black-T in 100 ml alcohol.

4. Preparation of buffer solution. Add 67.5 g of NH_4Cl to 570 ml of conc. ammonia solution (NH_4OH) and then dilute with distilled water to 1 litre.

5. Standardization of EDTA solution. Rinse and fill the burette with EDTA solution. Pipette out 50 ml of standard hard water in a conical flask. Add 10–15 ml of buffer solution and 4 to 5 drops indicator. Titrate with EDTA solution till wine-red colour changes to clear blue. Let volume used is V_1 ml.

6. Titration of unknown hard water. Titrate 50 ml of water sample just in step-5. Let volume used is V_2 ml.

7. Titration of permanent hardness. Take 250 ml of water sample in a large beaker. Boil it, till the volume is reduced to about 50 ml, filter it, wash the precipitate with distilled water, collecting filtrate and washing in a 250 ml measuring flask. Finally make up the volume to 250 ml with distilled water. Then, titrate 50 ml of boiled water sample just as in step (5). Let volume used is V_3 ml.

Calculations:

$$50 \text{ ml of standard hard water} = V_1 \text{ ml of EDTA}$$

$$\therefore \quad 50 \times 1 \text{ mg of } CaCO_3 = V_1 \text{ ml of the EDTA}$$

$$\therefore \quad 1 \text{ ml of EDTA} = \frac{50}{V_1} \text{ mg of } CaCO_3 \text{ eq}$$

$$\text{Now,} \quad 50 \text{ ml of given water} = V_2 \text{ ml of the EDTA} = \frac{V_2 \times 50}{V_1} \text{ mg of } CaCO_3 \text{ eq}$$

$$1 \text{ L (1000 ml) of given hard water} = \frac{1000 \times V_2}{V_1} \text{ mg of } CaCO_3 \text{ eq}$$

$$\text{Total hardness of water} = \frac{1000 \times V_2}{V_1} \text{ mg/L} = \frac{1000 \times V_2}{V_1} \text{ ppm}$$

$$\text{Now,} \quad 50 \text{ ml of boiled water} = V_3 \text{ ml of the EDTA} = \frac{V_3 \times 50}{V_1} \text{ mg of } CaCO_3 \text{ eq}$$

$$1 \text{ L (1000 ml) of boiled water} = \frac{1000 \times V_3}{V_1} \text{ mg of } CaCO_3 \text{ eq}$$

$$\text{Permanent hardness of water} = \frac{1000 \times V_3}{V_1} \text{ ppm}$$

$$\text{Temporary hardness} = \text{Total hardness} - \text{Permanent hardness}$$

$$= 1000\left[\frac{V_2}{V_1} - \frac{V_3}{V_1}\right] \text{ ppm} = \frac{1000 \times (V_2 - V_3)}{V_1} \text{ ppm}$$

Advantages of EDTA method: This method is definitely preferable to other methods, because of

(*i*) Greater accuracy,

(*ii*) Convenience, and

(*iii*) More rapid procedure.

1.6 POTABLE WATER OR DRINKING WATER OR MUNICIPAL WATER

Municipalities have to supply potable water; *Water which is safe to drink is known as potable water.*

Potable water fit for human consumption, should satisfy the following essential requirements:

1. It should be sparking clear and odourless.
2. It should be pleasant in taste.
3. It should be perfectly cool.
4. Its turbidity should not exceed 10 ppm.
5. It should be free from objectionable dissolved gases like hydrogen sulphide.
6. It should be free from objectionable minerals such as Pb, As, Cr and Manganese salts.
7. Its alkalinity should not be high. Its pH should be 8.0.
8. It should be reasonably soft.
9. Its total dissolved solids should be less than 500 ppm.
10. It should be free from disease-producing micro-organisms.

Purification of Water for Domestic use or Treatment of Water for Municipal Supply. Natural water from the rivers, canals, etc., does not confirm to all the required specifications of drinking water. For removing various types of impurities, the following methods are employed;

1. Removal of suspended impurities
2. Sterilisation or disinfection.

1. Removal of suspended impurities. Various impurities of water are removed in the following stages:

- **Screening.** The process of removing floating matter from the water is known as screening. In this process, water is passed through a screen, which arrests all the floating matter.
- **Aeration.** The water is subjected to aeration, which helps in
  - ➢ Exchange of gases,
  - ➢ Increased oxygen content, and
  - ➢ By removing the impurities like Fe and Mn by precipitating as their hydroxides.
- **Sedimentation.** It is a process of removing suspended impurities by allowing the water to stand undisturbed for 2-6 hours in a big tanks about 5 m deep, when most of the suspended particles settle down at the bottom, due to the force of gravity.

Sedimentation by coagulation. This is the process of removal of suspended colloidal impurities by using coagulants like

- ➢ Alum ($K_2SO_4 . Al_2(SO_4)_3 . 24H_2O$)
- ➢ Ferrous sulphate ($FeSO_4$)
- ➢ Sodium aluminate ($NaAlO_2$)

When a coagulant is added to water, floc formation takes place due to hydroxide [$Al(OH)_3$] formation which can gather tiny particles together to form bigger particles and settle down quickly.

$$Al_2(SO_4)_3 + 6H_2O \longrightarrow 2Al(OH)_3 + 3H_2SO_4$$

$$NaAlO_2 + 2H_2O \longrightarrow NaOH + Al(OH)_3$$

- **Filtration.** This process helps in removal of the colloidal and suspended impurities not removed by sedimentation. Usually the water is passed through filters or suitable porous material so as to remove the suspended impurities and some micro-organisms. Normally filters made of three layers like sand, gravel and charcoal is employed. Yet the Indian cotton sariee four fold will work as ultra filter. (Proved by the NASA testing, reported in News papers).

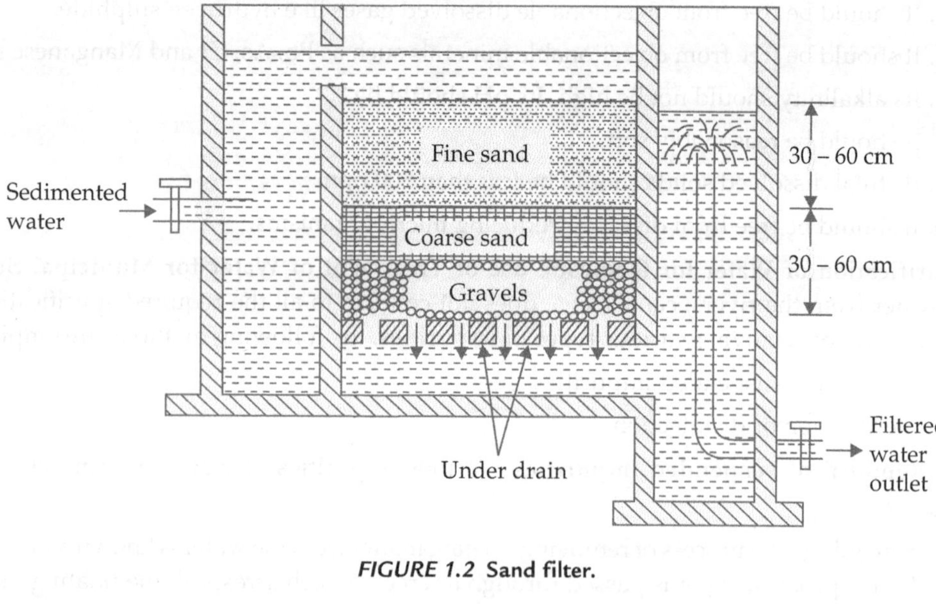

FIGURE 1.2 **Sand filter.**

1.7 STERILISATION AND DISINFECTION

The process of killing pathogens or other micro-organisms in the water is known as Sterilisation and Disinfection. The water which is free from micro-organisms or other pathogens is known as potable water.

Sterilisation is done in the following ways:

- **Boiling.** Boiling water for 15–20 minutes, will kill the bacteria in water. Yet this method is not suitable for municipal supply of water. This is adopted for domestic purpose.

- **Aeration or UV treatment.** Exposing water to UV rays or pumping water through nozzles into air in broad day light will kill the germs and micro-organisms in water.
- **Ozonisation.** Ozone passed into water acts as disinfectant. Ozone is an unstable isotope of oxygen, produces nascent oxygen which is a powerful disinfectant.

$$O_3 \longrightarrow O_2 + \underset{\text{Nascent oxygen}}{[O]}$$

This treatment is expensive and ozone cannot be stored for long time due to its instability.

- **Chemicals.** Chemicals like chlorine tablets and bleaching powder are more commonly used.
- **Chlorination.** Chlorine is also a powerful disinfectant. The use of chlorine for disinfecting water is known as chlorination. The chlorination is done in the following ways:
    - **By adding bleaching power.** Bleaching power contains 80% chlorine. When it is used as disinfectant, it is called hypo-chlorination because the disinfection is due to the formation of Hypochlorous acid formed during the process.

$$\underset{\text{Bleaching powder}}{CaOCl_2} + H_2O \longrightarrow Ca(OH)_2 + Cl_2$$

$$Cl_2 + H_2O \longrightarrow \underset{\text{Disinfectant}}{HOCl} + HCl$$

$$HOCl \longrightarrow [O] + HCl$$

   - **By passing chloramines.** Chlorine mixed with ammonia in the ratio of 2 : 1 by volume will produce a stable chloramine. This is a very powerful disinfectant. This is normally employed in swimming pools.

$$Cl_2 + NH_3 \longrightarrow NH_2Cl + HCl$$

$$NH_2Cl + H_2O \longrightarrow HOCl + NH_3$$

   - **Passing chlorine:** Directly administering the chlorine gas along with water in a chlorinator will be an effective process. The passed chlorine reacts with water and generates hypochlorous acid, which acts as disinfectant.

$$Cl_2 + H_2O \longrightarrow \underset{\text{Hypochlorous acid}}{HOCl} + HCl$$

Chlorine is a powerful disinfectant than chloramines and bleaching powder. Calculated amount must be added to water because in excess will be toxic and tastes bad. The amount of chlorine required to kill bacteria and to remove organic matter is called break point of chlorination.

Flow Diagram of Municipal Water Treatment. Municipal water treatment in short may be illustrated by flow diagram shown in Fig. 1.3.

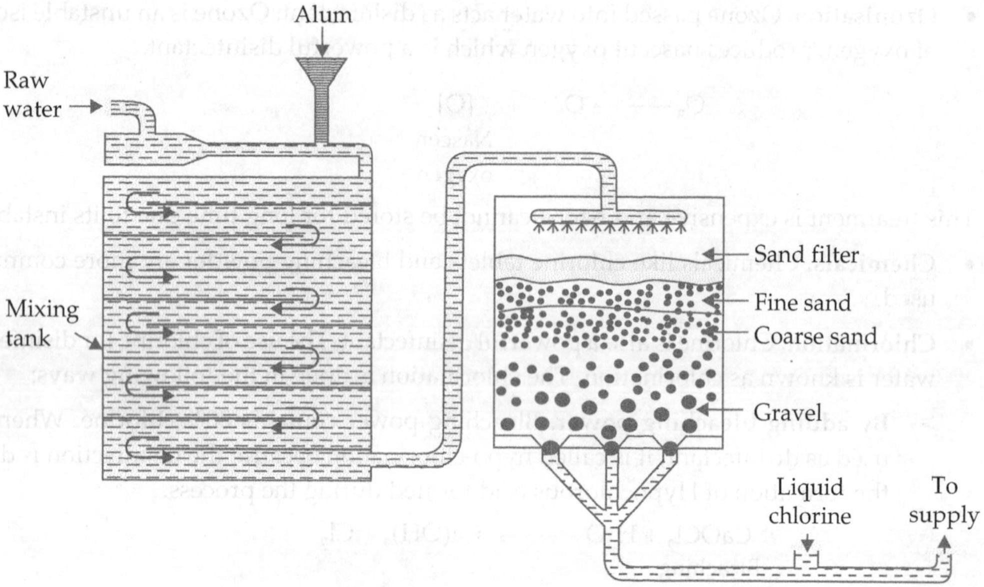

FIGURE 1.3 **Flow sheet of drinking water treatment plant.**

Water for industrial use. Water plays a key role in industry. Without it no industry runs. The major use is in production of steam and as a coolant. The water used for making steam should be free from dissolved salts, gases, oil, silica and suspended impurities. The steam is used in the production of electricity, as a heat source where temperature is an important criteria (like food processing plants) etc.

1.8 BOILER FEED WATER

The water fed into the boiler for the production of steam is called boiler feed water.

Requirements. Boiler feed water should be free from turbidity, oil, dissolved gases, alkali and hardness causing substances.

1.9 BOILER TROUBLES

In treatment of water complete elimination of all the impurities is not possible. The impurity that gives rise to certain troubles will be removed to certain extent. In modern pressure boilers and laboratories, water purer than the distilled water is required. Some of the boiler troubles caused by the use of unsuitable water are,

1. Priming and Foaming
2. Scale and Sludge formation
3. Boiler corrosion and
4. Caustic embrittlement.

1.9.1 Priming and Foaming

Priming. When a boiler is producing steam rapidly some particles of the water liquid carried along with the steam. This process of 'Wet Steam' formation is called priming.

Causes of priming:

1. The presence of large amount of dissolved solids.
2. High steam velocities.
3. Sudden boiling.
4. Improper boiler design.
5. Sudden increase in steam-production rate.

Disadvantages of priming:

1. Dissolved salt in boiler water are carried out by the wet steam to turbine blades which reduces their efficiency.
2. Dissolved salts may enter the parts of other machinery may decrease the life of machinery.
3. Actual height of the water column cannot be judged properly, thereby making the maintenance of the boiler pressure becomes difficult.

Prevention of priming:

1. Improving boiler design.
2. Fitting mechanical steam purifiers.
3. Maintaining low water level in boilers
4. Using soft water.
5. Decreasing the amount of dissolved salts.

Foaming. It is the production of foam or bubbles in boiler which do not break easily.

Causes of foaming. It is due to the presence of oily substances in water.

1. Low level of water in boiler.
2. The presence of dissolved salts in water.
3. Sudden increase in steam production rate.

Disadvantages of foaming:

1. Actual height of the water column cannot be judged.
2. Dissolved salts in water carried by the wet steam may damage turbine blades or machinery parts.
3. Boiler pressure cannot be maintained.

Prevention of foaming:

1. By the addition of anti-foaming chemicals like castor oil, Gallic acid, tennic acid etc.
2. Removing oil from boiler water by adding compounds like sodium aluminate.

1.9.2 Scale and Sludge Formation

In boilers, because of continuous evaporation of water, the concentration of salts increase progressively and after the saturation point is reached, precipitate forms on the inner walls of boiler.

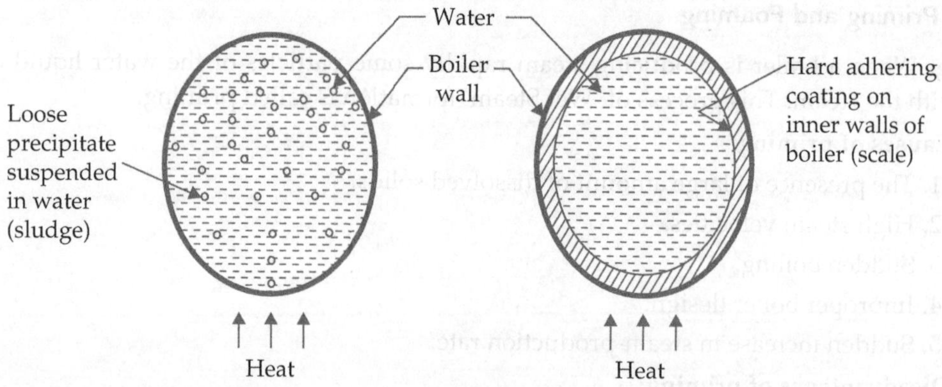

FIGURE 1.4

Sludge. Sludge is a soft, loose and slimy precipitate formed within the boiler. It is formed at comparatively colder portions of the boiler and collects in the area where flow rate is slow.

These are formed by substances which have greater solubilities in hot water than in cold water.

For example, $MgCO_3$, $MgCl_2$, $CaCl_2$, $MgSO_4$.

Disadvantages of sludge formation:

1. Sludges are poor conductors of heat, so they tend to waste a portion of heat generated.
2. Excessive sludge formation disturbs the working of boiler.
3. It causes choking of pipes.

Prevention of sludge formation:

1. By using soft water.
2. By frequently blow-down operation.
3. By removing the scale with the help of brush.
4. By giving thermal shocks. (*i.e.*, heating the boiler and then suddenly cooling with cold water).

Scales. Scales are hard, adhering precipitates formed on the inner walls of the boilers. They stick very firmly on to the inner wall surface and are difficult to remove with chisel and hammer. Due to scale formation, over heating of boiler is done in order to maintain a constant supply of steam. The overheating of the boiler tube makes the material softer and weaker and this causes distortion of boiler tube and also makes the boiler unsafe to bear the pressure of the steam.

Causes of Scale Formation. Following are the causes:

(*a*) **Decomposition of calcium bicarbonate.** Scale composed chiefly of calcium carbonate is soft and is the main cause of scale formation in low-pressure boilers.

$$Ca(HCO_3)_2 \longrightarrow \underset{\text{Scale}}{CaCO_3\downarrow} + H_2O + CO_2\uparrow$$

But in high pressure boilers, $CaCO_3$ becomes soluble.

$$CaCO_3 + H_2O \longrightarrow \underset{\text{Soluble}}{Ca(OH)_2} + CO_2\uparrow$$

(*b*) **Decomposition of calcium sulphate.** The solubility of $CaSO_4$ in water decreases with rise of temperature. In super heated water $CaSO_4$ is insoluble. This is the main cause in high pressure boilers.

(*c*) **Hydrolysis of magnesium salts.** Dissolved magnesium salts undergo hydrolysis forming $Mg(OH)_2$ precipitate.

$$MgCl_2 + 2H_2O \longrightarrow \underset{\text{Scale}}{Mg(OH)_2\downarrow} + 2HCl\uparrow$$

$Mg(OH)_2$ so found by hydrolysis of Magnesium salts is a soft type of scale.

(*d*) **Presence of silica.** Silica present in small quantities deposits as silicates like $CaSiO_3$ and $MgSiO_3$. These are very difficult to remove.

Disadvantages:

(*a*) **Wastage of fuel.** The scale formation causes decrease of heat transfer. As a result over heating is required this causes consumption of fuel.

(*b*) **Lowering of boiler safety.** Over-heating of boiler is done in order to maintain a steady supply of steam due to the formation of scales. This makes the boiler material softer and weaker. This causes the distortion of the boiler tube and also makes the boiler unsafe.

(*c*) **Danger of explosion.** The hot scale cracks because of expansion and water suddenly comes in contact with overheated Iron plates. This causes in formation of large amount of steam suddenly. This results high pressure causing boiler to burst.

(*d*) **Decrease in efficiency.** Scale may sometimes deposit in the valves and condensers of the boiler and choke them partially. This results in decrease in efficiency of boiler.

Removal of scales. If the scales are loosely adhering, they can be removed by scraper or piece of wood or wire brush.

1. If the scales are brittle, they can be removed by thermal shocks *i.e.*, heating the boiler and then suddenly cooling with cold water.
2. If the scales are loosely adhering, they can be removed by frequent blow-down operation.
3. If the scales are adherent and hard, they can be removed by adding chemicals.

Example. (*a*) $CaCO_3$ scales can be removed by dissolving in 5–10% HCl. (*b*) $CaSO_4$ scales can be removed by adding EDTA, since the Ca-EDTA complex is highly soluble in water.

1.9.3 Boiler Corrosion

The decay of boiler material by chemical or electrochemical attack of its environment is called boiler corrosion. The prime reasons of boiler corrosion are:

(*a*) Dissolved oxygen

(*b*) Dissolved carbon dioxide

(*c*) Action of acids in boiler.

(*a*) **Dissolved oxygen.** Among the dissolved gases oxygen is the most corroding impurity. At room temperature water contains 8 cc of oxygen per litre. The dissolved oxygen at high temperature attacks the boiler plate creating serious corrosion problem.

Removal of dissolved oxygen:

1. Addition of sodium sulphite or sodium sulphide removes O_2 by converting O_2 to sodium sulphate.

$$2Na_2SO_3 + O_2 \longrightarrow 2Na_2SO_4$$
$$Na_2S + 2O_2 \longrightarrow Na_2SO_4$$

2. Addition of hydrazine (NH_2—NH_2) is an ideal reagent to boiler to remove dissolved oxygen as H_2O.

$$NH_2NH_2 + O_2 \longrightarrow N_2 + 2H_2O$$

3. Mechanical de-aeration is another method of degasification. Water is sprayed through a perforated plate, fitted in the degasification tower, heated from sides and connected to vacuum pump as shown in Fig. 1.5. High temperature, low pressure and large exposed surface reduce dissolved oxygen and other gases in water.

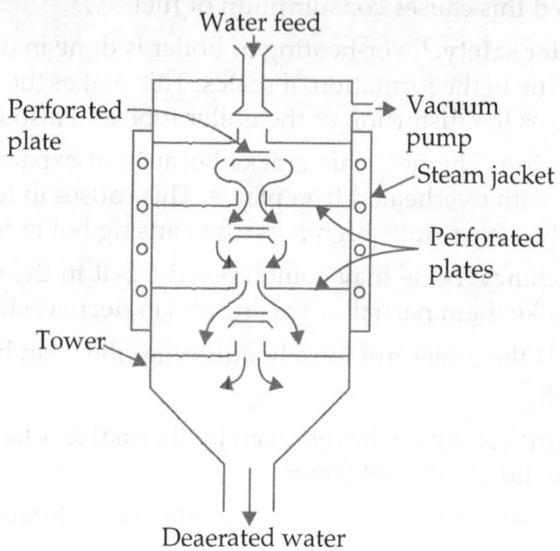

FIGURE 1.5

(*b*) **Dissolved CO_2.** Dissolved carbon dioxide has a slow corrosive effect on the materials of boiler plate. Source of carbon dioxide into water is the boiler feed water which contains bicarbonates. Under high temperature and pressure, maintained in the boiler the bicarbonates decompose to produce CO_2.

$$Ca(HCO_3)_2 \xrightarrow{\Delta} CaCO_3 + CO_2\uparrow + H_2O$$

$$Mg(HCO_3)_2 \xrightarrow{\Delta} Mg(OH)_2 + 2CO_2\uparrow$$

The carbon dioxide is removed by addition of calculated quantity of ammonia.

$$2NH_4OH + CO_2 \longrightarrow (NH_4)_2CO_3 + H_2O$$

Carbon dioxide is also removed by the using mechanical de-aerator.

(*c*) **Action of acids in boiler.** The acids are produced due to the hydrolysis of magnesium salts in the boiler.

$$MgCl_2 + 2H_2O \longrightarrow Mg(OH)_2\downarrow + 2HCl\uparrow$$

The acids produced react with iron of the boiler and decay the metal.

$$Fe + 2HCl \longrightarrow FeCl_2 + H_2\uparrow$$

$$FeCl_2 + 2H_2O \longrightarrow Fe(OH)_2 + 2HCl$$

$$2Fe(OH)_2 + H_2O + \frac{1}{2}O_2 \longrightarrow \underset{\text{Rust}}{[Fe_2O_3 . 3H_2O]}$$

Consequently even a small amount of $MgCl_2$ can cause corrosion to a large extent.

Preventive measures:

- Water softened before fed into boiler.
- Frequent blow–down of the boiler.
- Addition of inhibitors like sodium silicate, sodium phosphate and sodium chromate to slowdown corossion. The inhibitors form a thin film on the surface of the boiler and protect the metal from the attack of acid.

1.9.4 Caustic Embrittlement

Caustic embrittlement is a form of corrosion caused by high concentration of sodium hydroxide in the boiler water. It is characterized by the formation of irregular cracks on the boiler metal, particularly at places of high local stress, such as riveted seams, bends and joints. The caustic embrittlement occurs in boilers at high pressures, where NaOH is produced in the boiler by the hydrolysis of Na_2CO_3.

$$Na_2CO_3 + H_2O \longrightarrow 2NaOH + CO_2$$

The presence of sodium hydroxide makes the water alkaline. This alkaline water enters minute cracks present in the inner walls of the boiler by action. Inside the cracks, the water evaporates and amount of hydroxide keeps on increasing progressively. This sodium hydroxide attacks the surrounding material and dissolves iron of the boiler as sodium ferroate (Na_2FeO_2).

This causes embrittlement of boiler parts like rivets, bends and joints, which are under stress.

$$\underset{\text{Insoluble}}{Fe} + 2NaOH \longrightarrow \underset{\text{Soluble}}{Na_2FeO_2} + H_2$$

Prevention of Caustic Embrittlement:

1. By using sodium phosphate as softening reagent instead of sodium carbonate. Di sodium hydrogen phosphate is the best softening reagent because it not only forms complex with Ca^{2+} and Mg^{2+} resulting the softening of water but also maintains pH of water at 9–10. The other phosphates used are tri sodium phosphate and sodium di hydrogen phosphate etc.
2. By adding tannin or lignin to the boiler water which block the hair cracks and pits that are present on the surface of the boiler plate thus preventing the infiltration of caustic soda solution.
3. By adding sodium sulphate to boiler water, which also blocks the hair cracks and pits present on the surface of the boiler plate, preventing the infiltration of the caustic soda solution.

1.10 TURBINE DEPOSITS

Although several factors influence the formation of deposits on turbine components, the general effect is the same no matter what the cause. Adherent deposits form in the steam passage, distorting the original shape of turbine nozzles and blades. These deposits, often rough or uneven at the surface, increase resistance to the flow of steam. Distortion of steam passages alters steam velocities and pressure drops, reducing the capacity and efficiency of the turbine. Where conditions are severe, deposits can cause excessive rotor thrust. Uneven deposition can imbalance the turbine rotor, causing vibration problems.

As deposits accumulate on turbine blades, stage pressures increase. The deposits were caused by the use of contaminated water to temperate the steam. In a fouled condition, this 30 MW turbine lost over 5% of its generating capacity.

Turbine deposits can accumulate in a very short time when steam purity is poor. The turbine was forced off-line by deposition only 3 months after it was placed in operation. Carryover of boiler water, resulting from inadequate steam-water separation equipment in the boiler, caused this turbine deposit problem.

1.11 SOFTENING OF WATER

Definition. The process of removing hardness-causing salts from water is known as "Softening of water".

Internal Treatment. The internal treatment consists of adding chemicals directly to the water in the boilers for removing scales and sludges.

1. **Colloidal conditioning.** Organic substances like kerosene and agar-agar are added to water which prevent the scales.
2. **Phosphate conditioning.** Phosphates react with water and a loose sludge is formed which can be easily removed by blow-down operation.

$$3CaSO_4 + 2Na_3PO_4 \longrightarrow Ca_3(PO_4)_2 + 3Na_2SO_4$$

3. **Carbonate conditioning.** In low pressure boilers, scale formation can be avoided by adding sodium carbonate to the boiler water.

$$CaSO_4 + Na_2CO_3 \rightleftharpoons CaCO_3 + Na_2SO_4$$

4. **Calgon conditioning.** Sodium hexa meta phosphate (calgon) is added to boiler water to prevent scale formation by formation of soluble complex.

$$2CaSO_4 + Na_2(Na_4(PO_3)_6) \longrightarrow Na_2(Ca_2(PO_3)_6) + 2Na_2SO_4$$

5. **Sodium aluminate conditioning.** Sodium aluminate gets hydrolysed yielding NaOH and aluminium hydroxide precipitate. The sodium hydroxide reacts with magnesium salts to magnesium hydroxide. These hydroxides entrap finely divided particles and neutralize the charge on colloidal particles, and the loose precipitate can be removed by the blow-down operation.

$$NaAlO_2 + 2H_2O \longrightarrow NaOH + Al(OH)_3\downarrow$$

Softening of Water by External Treatment. Softening of water means the removal of calcium, magnesium, iron salts and similar other metallic ions, which would form insoluble metallic soaps. The three important industrial methods employed for softening of water are:

1. Lime-Soda process.
2. Permutit or Zeolite process.
3. Ion Exchange or Demineralization process.

1.11.1 Lime-Soda Process

This is the most important and economical treatment of water softening. The principle involved is, to chemically convert all the soluble hardness-causing substances into insoluble precipitates and latter removed by settling and filtering. A calculated amount of lime and soda along with coagulants are added to the water for softening. The reactions can be summerised as follows.

- Lime removes the temporary hardness

$$Ca(OH)_2 + Ca(HCO_3)_2 \longrightarrow 2CaCO_3\downarrow + 2H_2O$$
$$2Ca(OH)_2 + Mg(HCO_3)_2 \longrightarrow Mg(OH)_2\downarrow + 2CaCO_3\downarrow + 2H_2O$$

- Lime removes the permanent magnesium hardness

$$Ca(OH)_2 + MgCl_2 \longrightarrow Mg(OH)_2\downarrow + CaCl_2$$
$$Ca(OH)_2 + MgSO_4 \longrightarrow Mg(OH)_2\downarrow + CaSO_4$$

- Lime removes dissolved iron and aluminium salts

$$Ca(OH)_2 + FeSO_4 \longrightarrow Fe(OH)_2\downarrow + CaSO_4$$
$$2Fe(OH)_2 + H_2O + [O] \longrightarrow 2Fe(OH)_3\downarrow$$
$$3Ca(OH)_2 + Al_2(SO_4)_3 \longrightarrow 2Al(OH)_3\downarrow + 3CaSO_4$$

- Lime removes free mineral acids

$$Ca(OH)_2 + 2HCl \longrightarrow CaCl_2\downarrow + 2H_2O$$
$$Ca(OH)_2 + H_2SO_4 \longrightarrow CaSO_4 + 2H_2O$$

- Lime removes CO_2 and H_2S

$$Ca(OH)_2 + CO_2 \longrightarrow CaCO_3\downarrow + 2H_2O$$
$$Ca(OH)_2 + H_2S \longrightarrow CaS\downarrow + 2H_2O$$

- Lime removes sodium and potassium bicarbonates

$$2NaHCO_3 + Ca(OH)_2 \longrightarrow CaCO_3 + 2H_2O + Na_2CO_3$$
$$2KHCO_3 + Ca(OH)_2 \longrightarrow CaCO_3 + 2H_2O + K_2CO_3$$

- Soda removes all soluble calcium permanent hardness

$$Na_2CO_3 + CaCl_2 \longrightarrow CaCO_3\downarrow + 2NaCl$$
$$Na_2CO_3 + CaSO_4 \longrightarrow CaCO_3\downarrow + Na_2SO_4\downarrow$$

Calculation of Lime-Soda Required for Softening. From the above said reactions it is evident that

- One equivalent of calcium temporary hardness requires one equivalent of lime. [From 1]
- One equivalent of magnesium temporary hardness requires two equivalents of lime. [From 1]
- One equivalent of calcium permanent hardness requires one equivalent of soda. [From 6]

- One equivalent of magnesium permanent hardness requires one equivalent of lime and one equivalent of soda

$$MgSO_4 + Ca(OH)_2 + Na_2CO_3 \longrightarrow Mg(OH)_2 + CaCO_3 + Na_2SO_4$$
$$MgCl_2 + Ca(OH)_2 + Na_2CO_3 \longrightarrow Mg(OH)_2 + CaCO_3 + 2NaCl$$

- Lime while reacting with HCl, H_2SO_4, $MgSO_4$, $MgCl_2$, $Mg(NO_3)_2$, salts of Fe, Al etc. generates the corresponding quantities of calcium permanent hardness. Accordingly these constituents also should be considered while calculating soda requirements.

Impurity	*Multiplicative factor*
$Al_2(SO_4)_3$	3
HCl	0.5
$Mg(HCO_3)_2$	2
$NaHCO_3$, $KHCO_3$, $NaAlO_2$	0.5
NaCl, KCl, Silica, K_2SO_4, Fe_2O_3	Not considered

- Two equivalents of HCO_3^- reacts with one equivalents of lime.

$$2HCO_3^- + Ca(OH)_2 \longrightarrow CaCO_3 + 2H_2O + CO_3^{2-}$$

It is evident that in the above reaction, equivalent of CO_3^{2-} is generated. Thus for every one equivalent of HCO_3^- present, the corresponding reduction in the dose of soda has to be made in the calculations for soda requirement.

Steps to be followed for Calculating the Lime-Soda Requirement:

1. The units in which the impurities are analyzed should be one (ppm or mg/L or °C etc.).
2. The substances which do not contribute to hardness like KCl, NaCl, SiO_2, Na_2SO_4, Fe_2SO_3, K_2SO_4, etc. can be ignored from the calculations. This fact should be explicitly stated.
3. All the substances causing hardness should be converted into their respective $CaCO_3$ equivalents.
4. Lime soda requirement is calculated as follows.

Amount of lime required for softening

$$= \frac{74}{100}\left(\begin{array}{l}\text{temp. } Ca^{2+} \text{ hardness} + (2 \times \text{temp. } Mg^{2+} \text{ hardness}) + \text{permanent } Mg^{2+} \text{ hardness} \\ + CO_2 + \frac{1}{2}HCl + H_2SO_4 + \frac{1}{2}NaHCO_3 + \frac{1}{2}KHCO_3 + FeSO_4 + (3 \times Al_2(SO_4)_3) \\ -\left(\frac{1}{2}NaAlO_2\right) - \text{all the hardness in terms of } CaCO_3 \text{ equivalents.}\end{array}\right)$$

Amount of soda required for softening

$$= \frac{106}{100}\left(\begin{array}{l}\text{permanent } Ca^{2+} \text{ hardness} + \text{permanent } Mg^{2+} \text{ hardness} + \frac{1}{2}HCl \\ + H_2SO_4 + (3 \times Al_2(SO_4)_3) + FeSO_4 - \left(\frac{1}{2}KHCO_3\right) - \left(\frac{1}{2}NaHCO_3\right) \\ - \text{All in terms of } CaCO_3 \text{ equivalents.}\end{array}\right)$$

Notes:

1. The multiplicative factors 74/100 and 106/100 is taken since 100 parts by weight of $CaCO_3$= 74 parts of lime and 106 parts of soda.
2. If the impurities are given as $CaCO_3$ and $MgCO_3$, they should be considered as $Ca(HCO_3)_2$ and $Mg(HCO_3)_2$ respectively.
3. If the lime and soda are impure and if the percentage of purity is given, then the actual requirements of chemicals should be calculated accordingly by having multiplicative factor $\frac{100}{\text{\% of impurity}}$ For example, if the impurity is 85%, the multiplicative factor 100/85

Lime-soda process is of two types:

(*a*) Cold lime-soda process

(*b*) Hot lime-soda process

(*a*) **Cold lime-soda process.** In this method, calculated quantity of chemicals and water, along with accelerators and coagulators are added to a tank fitted with a stirrer shown in Fig. 1.6. On vigorous stirring, through mixing takes place. After softening the soft water rises upwards and the heavy sludges settle down. The softened water passes through a filtering media ensuring complete removal of the sludge and finally the filtered water flows out through the top. Cold lime soda process is used for partial softening of municipal water, for softening of cooling water etc. In actual purpose, magnesium hardness is brought down to almost zero but calcium hardness remains about 40 ppm.

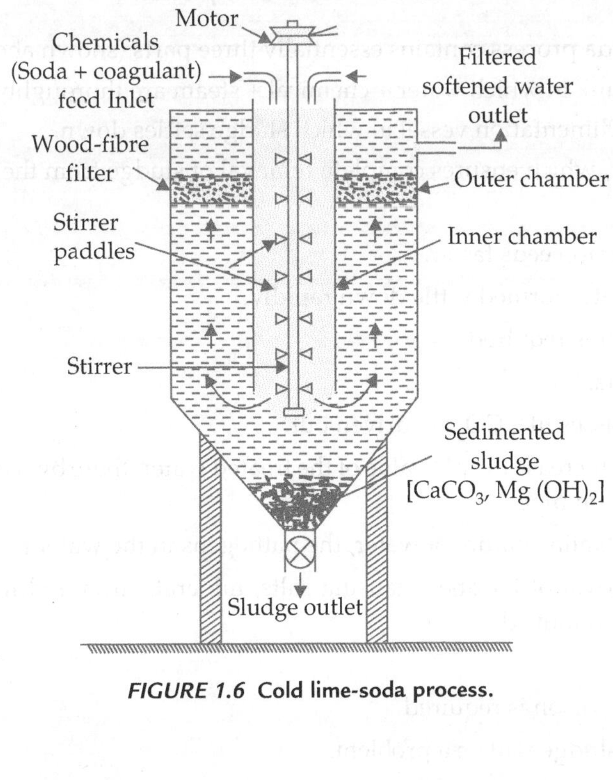

FIGURE 1.6 **Cold lime-soda process.**

(*b*) **Hot lime-soda process.** In this process the raw water is treated with softening chemicals at 80 – 150°C. This method is better than cold lime-soda process.

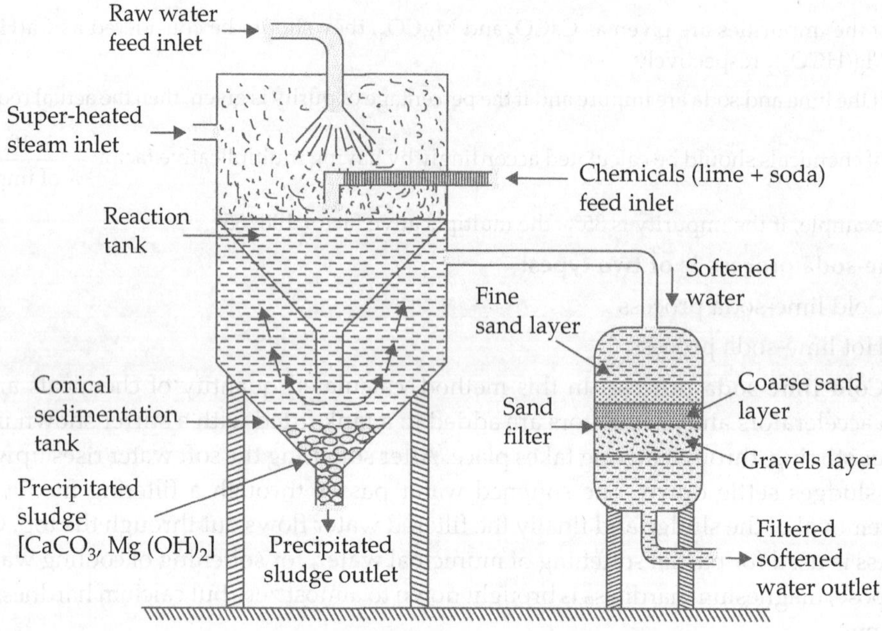

FIGURE 1.7 **Hot lime-soda process.**

The hot lime-soda process contains essentially three parts (shown above).

1. A reaction tank in which water + chemical + steam are thoroughly mixed.
2. A conical sedimentation vessel in which sludge settles down.
3. A sand filter, which ensures complete removal of sludge from the softened water.

Advantages:

- The reaction proceeds faster.
- The precipitates formed settle down rapidly.
- No coagulant is required.
- Filtration is fast.
- Dissolved gasses like CO_2 are driven out.
- This process increase the pH value of the treated water, there by reduces the corrosion of the distribution pipes.
- Due to the alkaline nature of water, the pathogens in the water are also reduced.
- Besides removal of hardness causing salts, minerals, iron and manganese present in water is also removed.

Disadvantages:

- Skilled supervision is required.
- Disposal of sludge will be a problem.

- The softened water contains 15–30 ppm of residual hardness which is not good for boilers.

Advantages of lime-soda process:

1. Lime-soda process is economical.
2. The process improves corrosion resistance of the water.
3. Mineral content of the water is reduced.
4. pH of the water rises, which reduces the content of pathogenic bacteria.

Disadvantages of lime-soda process:

1. Huge amount of sludge is formed, hence disposal is difficult.
2. Due to residual hardness, water is not suitable for high pressure boiler.

1.11.2 Permutit or Zeolite Process

Zeolite is hydrated sodium alumino silicate capable of exchanging reversibly its sodium ions for Ca^{2+} and Mg^{2+}, having the general formula $Na_2O\ .\ Al_2O_3 . xSiO_2 . yH_2O$. Common zeolite is $Na_2O . Al_2O_3 . 3SiO_2 . 2H_2O$ and is known as natrolith. Others gluconites, green sand etc. are used for water softening. Artificial zeolite used for softening purpose is permutit. These are porous and glassy and have greater softening capacity than green sand. They are prepared by heating together with china clay, feldspar and soda ash.

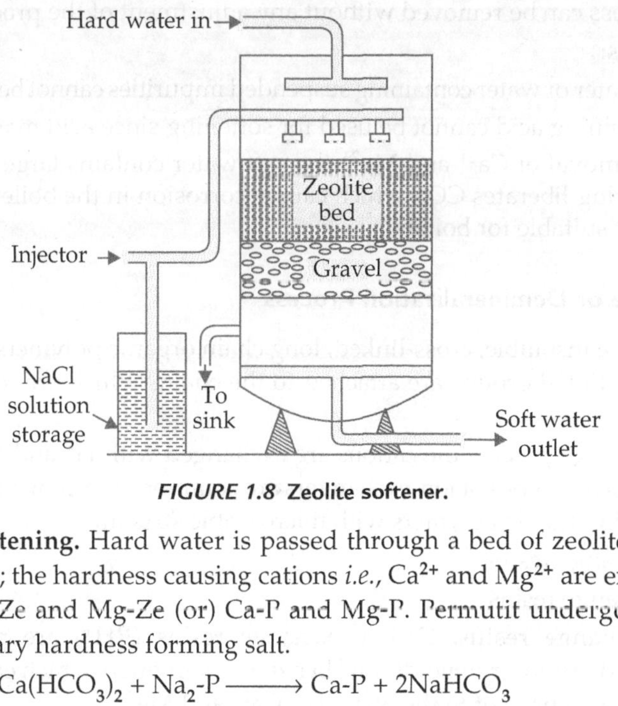

FIGURE 1.8 **Zeolite softener.**

Method of softening. Hard water is passed through a bed of zeolite at a specific rate at ordinary temperature; the hardness causing cations *i.e.*, Ca^{2+} and Mg^{2+} are exchanged for Na and it is converted to Ca-Ze and Mg-Ze (or) Ca-P and Mg-P. Permutit undergo following chemical reaction with temporary hardness forming salt.

$$Ca(HCO_3)_2 + Na_2\text{-}P \longrightarrow Ca\text{-}P + 2NaHCO_3$$

$$Mg(HCO_3)_2 + Na_2\text{-}P \longrightarrow Mg\text{-}P + 2NaHCO_3$$

Permutit undergo following chemical reaction with permanent hardness forming salt.

$$CaCl_2 + Na_2\text{-}P \longrightarrow Ca\text{-}P + 2NaCl$$

$$MgCl_2 + Na_2\text{-}P \longrightarrow Mg\text{-}P + 2NaCl$$

$$CaSO_4 + Na_2\text{-}P \longrightarrow Ca\text{-}P + Na_2SO_4$$
$$MgSO_4 + Na_2\text{-}P \longrightarrow Mg\text{-}P + Na_2SO_4$$

When permutit is completely converted into calcium and magnesium permutit and it ceases to soften the water. It gets exhausted. At this stage, the supply of hard water is stopped and the exhausted permutit is regenerated by treating the bed with concentrated 10% NaCl solution.

Regeneration of zeolite. The process is also commercially successful since the Ca/Mg zeolites formed by passing hard water through the bed can be easily regenerated into Na_2Ze by passing brine (10% NaCl) through the bed of inactivated zeolite.

$$Ca\text{-}P + 2NaCl \longrightarrow Na_2\text{-}P + CaCl_2$$
$$Mg\text{-}P + 2NaCl \longrightarrow Na_2\text{-}P + MgCl_2$$

Washing with water is led to remove $CaCl_2$ and $MgCl_2$ and thus obtained regenerated permutit can be again used.

Advantages:

1. Hardness of water can be removed completely up to about 10 ppm;
2. The equipment used is small and easy to handle;
3. It requires less time for softening;
4. There is no sludge formation, hence the process is clean;
5. Easy to regenerate;
6. Any hardness can be removed without any adjustment of the process.

Disadvantages:

1. Coloured water or water containing suspended impurities cannot be used before filtration;
2. Water containing acid cannot be used for softening since acid may destroy the Zeolite;
3. Since on removal of Ca^{2+} and Mg^{2+} the soft water contains large amount of $NaHCO_3$, this on heating liberates CO_2, which causes corrosion in the boilers and hence this soft water is not suitable for boilers.

1.11.3 Ion Exchange or Demineralization Process

Ion exchange resins are insoluble, cross-linked, long chain organic polymers with a micro porous structure and the functional groups are attached to the chains which are responsible for the ion exchange properties.

In the ion exchange process the cations are exchanged with H^+ and anions are exchanged with OH^- ions. For this two types of ion-exchangers are commonly employed, which are insoluble, cross-linked long chain organic polymers with microscopic structure.

(*i*) Cation exchange resins

(*ii*) Anion exchange resins

(*i*) **Cation exchange resins.** Cation exchange resins (RH^+) are phenol-sulfonic acid-formaldehyde resin, styrene-divinyl benzene copolymers which exchange their H^+ ions with the cations present in the water *i.e.*, Ca^{2+} and Mg^{2+}.

(*ii*) **Anion exchange resins.** Anion exchange resins (ROH^-) are the styrene divinyl benzene or amine formaldehyde copolymers contain quaternary ammonium tertiary sulphonium or amino group in the resin. The resin on treatment with NaOH solution is capable of exchanging the OH^- with different anions of water. *i.e.*, Cl^-, SO_4^{2-} etc.

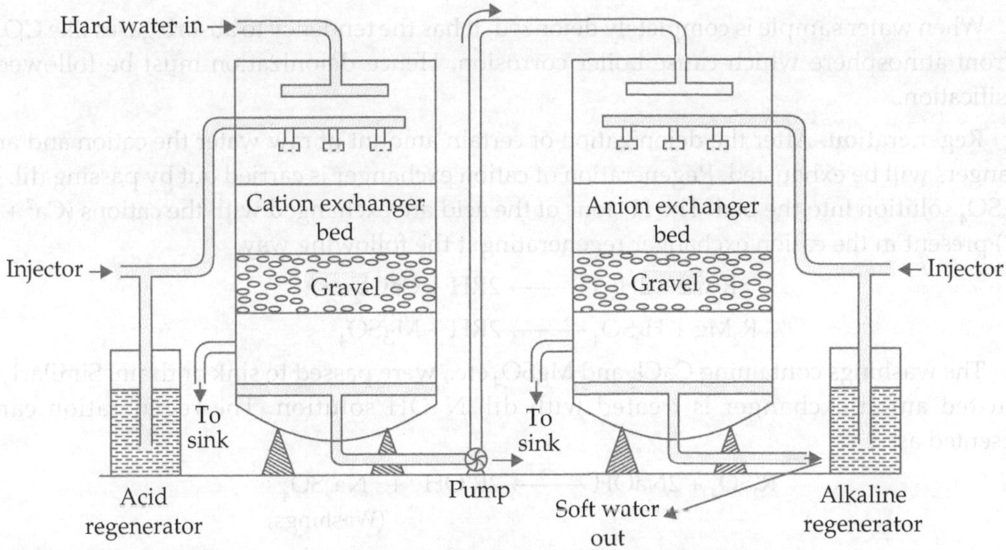

FIGURE 1.9 **Demineralization of water.**

Method. The hard water is passed first through cation exchange resin similar to the permutit process where the cations like Ca^{2+}, Mg^{2+} are removed from the hard water and exchanged with H^+ as follows,

$$2RH + Ca(HCO_3)_2 \longrightarrow R_2Ca + 2H_2CO_3$$
$$2RH + Mg(HCO_3)_2 \longrightarrow R_2Mg + 2H_2CO_3$$
$$2RH + CaCl_2 \longrightarrow R_2Ca + 2HCl$$
$$2RH + CaSO_4 \longrightarrow R_2Ca + H_2SO_4$$
$$2RH + Ca(NO_3)_2 \longrightarrow R_2Ca + 2HNO_3$$
$$2RH + MgCl_2 \longrightarrow R_2Mg + 2HCl$$
$$2RH + MgSO_4 \longrightarrow R_2Mg + H_2SO_4$$
$$2RH + Mg(NO_3)_2 \longrightarrow R_2Mg + 2HNO_3$$

The Ca^{2+} and Mg^{2+} are retained by the cation exchangers as Ca-R and Mg-R releasing H^+ into water. Thus the water coming out of the resin is highly acidic because the H^+ released by the exchange combines with anion of the dissolved salt to produce the corresponding acids.

After this the hard water is again passed through anion exchange column, which exchanges all the anions like SO_4^{2-}, Cl^- etc. present in the water with OH^-. The H^+ and OH^- released from exchangers get combined and produce H_2O.

$$R'OH + HCl \longrightarrow R'Cl + H_2O$$
$$2R'OH + H_2SO_4 \longrightarrow R_2'SO_4 + 2H_2O$$
$$R'OH + H_2CO_3 \longrightarrow R'HCO_3 + H_2O$$
$$R'OH + HNO_3 \longrightarrow R'NO_3 + H_2O$$

Thus the water coming out finally from the two exchangers is ion free and called deionized or demineralized water.

When water sample is completely deionzed, it has the tendency to absorb gases like CO_2, O_2 etc, from atmosphere which cause boiler corrosion. Hence deionization must be followed by degasification.

Regeneration. After the deionization of certain amount of raw water the cation and anion exchangers will be exhausted. Regeneration of cation exchanger is carried out by passing dil. HCl or H_2SO_4 solution into the bed. The H^+ ions of the acid are exchanged with the cations (Ca^{2+} and Mg^{2+}) present in the cation exchanger regenerating it the following way.

$$R_2Ca + 2HCl \longrightarrow 2RH + CaCl_2$$

$$R_2Mg + H_2SO_4 \longrightarrow 2RH + MgSO_4$$

The washings containing $CaCl_2$ and $MgSO_4$ etc., were passed to sink or drain. Similarly the exhausted anion exchanger is treated with dil. NaOH solution. The regeneration can be represented as;

$$R'_2SO_4 + 2NaOH \longrightarrow 2R'OH + \underset{\text{(Washings)}}{Na_2SO_4}$$

$$R'Cl + NaOH \longrightarrow R'OH + \underset{\text{(Washings)}}{NaCl}$$

$$R'HCO_3 + NaOH \longrightarrow R'OH + \underset{\text{(Washings)}}{NaHCO_3}$$

$$R'NO_3 + NaOH \longrightarrow R'OH + \underset{\text{(Washings)}}{NaNO_3}$$

The washings are discarded into sink. The regenerated ion-exchangers are used for softening. Thus deionization and regeneration are the alternate process.

Advantages:

1. Highly acidic or alkaline water samples can be purified by this process.
2. The hardness possessed by the deionized water is 2 ppm.
3. The deionized water is most suitable for high pressure boilers.

Disadvantages:

1. The ion exchanging resins are expensive; hence the cost of purification is high.
2. Raw water should contain turbidity below 10 ppm. Otherwise pores in the resin will be blocked and output of the process is reduced.

1.12 DESALINATION

"The process of removing common salt (NaCl) from the water is known as ***Desalination***".

The water containing dissolved salts with a peculiar salty (or brackish) taste is called brackish water. Sea water containing on an average about 3.5% salts, comes under this category. Brackish water is totally unfit for drinking purpose. Commonly used methods for the desalination of brackish water are

1. Osmosis,
2. Reverse Osmosis.

1.13 OSMOSIS

Principle. When two solutions of unequal concentrations are separated by a semi-permeable membrane, the pressure applied for the flowing of the solvent from dilute to concentrated side is known as osmotic pressure and the process is known as *Osmosis*. This is the principle of osmosis.

Example. Electrodialysis.

Electrodialysis. Electrodialysis is a method in which the ions (of the salt present) are pulled out of the salt water by passing direct current, using electrodes and thin rigid plastic membrane pair (Natural or Synthetic). The Fig. 1.10 illustrates the method of desalination by electrodialysis.

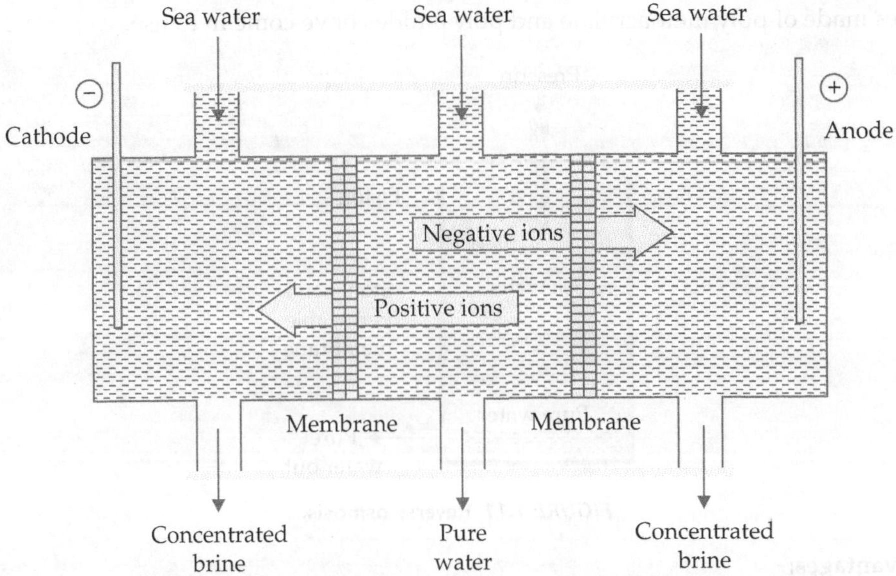

FIGURE 1.10 **Electrodialysis.**

When direct electric current is passed through saline water, the sodium ions (Na^+) start moving towards negative pole (cathode), while the chloride ions (Cl^-) start moving towards the positive pole (anode) through the membrane. As a result the concentration of brine decreases in the central compartment, while it increases in two side compartments. Desalinated brine (or pure water) is removed from the central compartment from time to time, while concentrated brine (inside the compartments) is replaced by fresh brine or sea water.

Advantages:

1. It is most compact unit.
2. The cost of installation of the plant and its operation is economical.
3. If electricity is easily available, it is best suited.

1.14 REVERSE OSMOSIS

Principle. When two solutions of unequal concentrations are separated by a semi-permeable membrane, flow of solvent takes place from dilute to concentrated sides, due to osmosis. If however, a hydrostatic pressure in excess of osmotic pressure is applied on the concentrated side, the solvent flow reverse, *i.e.*, solvent is forced to move from concentrated side to dilute side across the membrane. This is the ***principle of reverse osmosis***. Thus in reverse osmosis methods, pure solvent (water) is separated from its contaminates, rather than removing contaminants from the water. This membrane filtration is sometimes also called ***"Super- filtration"*** or ***"Hyper-filtration"***.

Method. In this process, pressure (of the order 15 to 40 kg/cm^2) is applied to the Sea water/impure water to force its pure water out through the semi-permeable membrane, leaving behind the dissolved salts (both ionic as well as non ionic). The principle of reverse osmosis, as applied for treating saline/Sea water, is illustrated in Fig. 1.11. The membrane consists of very thin films of cellulose acetate, affixed to either side of a perforated tube. However, more recently superior membranes made of poly methacrylate and poly amides have come in to use.

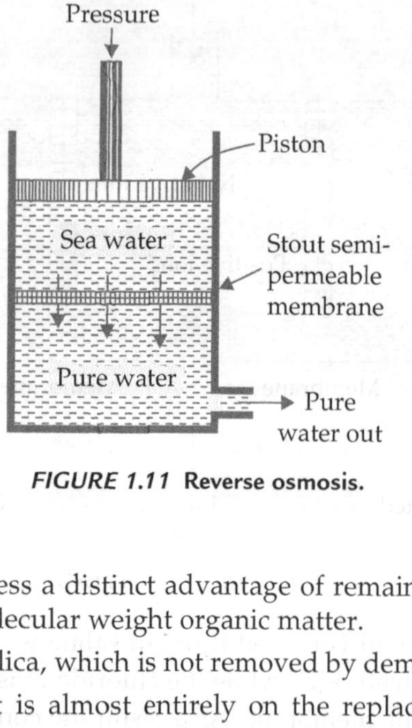

FIGURE 1.11 **Reverse osmosis.**

Advantages:

1. Reverse osmosis process a distinct advantage of remaining ionic as well as non-ionic, colloidal and high molecular weight organic matter.
2. It removes colloidal silica, which is not removed by demineralization.
3. The maintenance cost is almost entirely on the replacement of the semi-permeable membrane.
4. The life time of membrane is quite high, about 2 years.
5. The membrane can be replaced within few minutes, thereby it provides nearly uninterrupted water supply.
6. Due to low capital cost, simplicity, low operating cost and high reliability, the reverse osmosis is gaining ground at present for converting sea water into drinking water and for obtaining water for very high pressure boilers.

1.15 TYPES OF MEMBRANES USED IN DESALINATION PROCESS

Natural membranes like, pig's bladder, skin around egg white, membrane around the red blood capsules and the membrane of plant's cell are semi-permeable membranes. Animal and vegetable membranes are not completely semi-permeable.

$[Cu_2Fe(CN)_6]$ membrane deposited on the walls of a porous pot is perfectly semi-permeable and is used for accurate experimental work. All semi-permeable membranes have fine holes or capillaries in their structure. They allow passage of solvent molecules but not the larger solute molecules.

An ion selective membrane has permeability for only one kind of ions with specific charge. For example a cation-selective membrane is permeable to cations, due to negatively charged fixed functional groups. Similarly anion-selective membrane has positively charged fixed functional groups, and also polymers like polymethacrylate and poly amides and thin films of cellulose acetate used as a membrane in desalination process.

Advantages and disadvantages of desalination (Limitations):

1. It costs too much to pay for the plants.
2. The ocean is always there and easy to get feed water.
3. Make more green house gas so there for more pollution in the atmosphere.
4. It doubles the amount of salt in the ocean's and impacts on the environment for animals and humans.

SOLVED PROBLEMS

Problem 1. *Calculate the temporary and permanent hardness of sample of water containing: $Mg(HCO_3)_2$ = 7.3 mg/L; $Ca(HCO_3)_2$ = 16.2 mg/L; $MgCl_2$ = 9.5 mg/L; $CaSO_4$ = 13.6 mg/L. (Atomic weights of Mg and Ca are 20 and 40 respectively.)*

Solution. Conversion into $CaCO_3$ equivalents:

Costituent	*Multiplication factor*	*$CaCO_3$ equivalent*
$Mg(HCO_3)_2$ = 7.3 mg/L	100/146	7.3 × 100/146 = 5 mg/L
$Ca(HCO_3)_2$ = 16.2 mg/L	100/162	16.2 × 100/162 = 10 mg/L
$MgCl_2$ = 9.5 mg/L	100/95	9.5 × 100/95 = 10 mg/L
$CaSO_4$ = 13.6 mg/L	100/136	13.6 × 100/136 = 10 mg/L

Temporary hardness, due to $Mg(HCO_3)_2$ and $Ca(HCO_3)_2$

= (5 +10) mg/L = 15 mg/L or 15 ppm

Permanent hardness, due to $MgCl_2$ and $CaSO_4$

= (10 + 10) mg/L = 20 mg/L or 20 ppm

Total hardness of given sample of water =Temporary hardness + Permanent hardness

= (15 + 20) ppm = 35 ppm.

Problem 2. *50 ml of a sample water consumed 15 ml of 0.01 M EDTA before boiling and 5 ml of the same EDTA after boiling. Calculate the degree of total hardness, permanent hardness and temporary hardness.*

Solution. 50 ml of water sample = 15 ml of 0.01 M EDTA

1 litre of water sample (total hardness)

$$= \frac{15 \times 1000}{50} \text{ ml of } 0.01 \text{ M EDTA}$$

= 300 ml of 0.01 M EDTA

= 2 × 300 ml of 0.01 N EDTA

(∵ Molarity of EDTA = 2 × Normality of EDTA)

= 600 ml (or 0.6 L) of 0.01 eq $CaCO_3$

= 0.6 × 0.01 × 50 g $CaCO_3$ eq

Hence, the total hardness = 0.30 g or 300 mg of $CaCO_3$ eq

= 300 mg/L or 300 ppm.

Now, 50 ml of boiled water = 5 ml of 0.01 M EDTA

$$\therefore \quad 1000 \text{ ml of boiled water} = \frac{5 \times 1000}{50} \text{ ml of } 0.01 \text{ M EDTA}$$

= 100 ml of 0.01 M EDTA

= 2 × 100 ml of 0.01 N EDTA

(∵ Molarity of EDTA = 2 × Normality of EDTA)

= 200 ml (or 0.2 L) of 0.01 N EDTA

= 0.2 × 0.01 × 50 g $CaCO_3$ eq

Hence, the permanent hardness = 0.1 g or 100 mg of $CaCO_3$ eq

= 100 mg/L or 100 ppm.

∴ Temporary hardness = (Total hardness – Permanent hardness)

= (300 – 100) ppm = 200 ppm.

Problem 3. *Calculate the quantity of Lime and Soda required for softening 50,000 litres of water containing the following salts for litre: $Mg(HCO_3)_2$ = 7.5 mg; $Ca(HCO_3)_2$ = 8.1 mg; $CaSO_4$ = 13.6 mg; $MgSO_4$ = 12.0 mg; $MgCl_2$ = 2.0 mg; NaCl = 4.7 mg.*

Solution. Calculation of $CaCO_3$ equivalents:

Costituent	*Multiplication factor*	*$CaCO_3$ equivalent*
$Ca(HCO_3)_2$ = 8.1 mg/L	100/162	8.1 × 100/162 = 5 mg/L
$Mg(HCO_3)_2$ = 7.5 mg/L	100/146	7.5 × 100/146 = 5.14 mg/L
$CaSO_4$ = 13.6 mg/L	100/136	13.6 × 100/136 = 10 mg/L
$MgSO_4$ = 12.0 mg/L	100/120	12.0 × 100/120 = 10 mg/L
$MgCl_2$ = 2.0 mg/L	100/95	2 × 100/95 = 2.11 mg/L

Lime requirement for 50,000 L water

$$= \frac{74}{100} [Ca(HCO_3)_2 + 2\,Mg(HCO_3)_2 + MgSO_4 + MgCl_2 \text{ as } CaCO_3 \text{ eq}] \times \text{Volume of water}$$

$= 74/100[5.0 + 2 \times 5.14 + 10.0 + 2.11]$ mg/L × 50,000 L

$= 74/100[27.39]$ mg/L × 50,000 L = 10,13,430 mg

$= 10,13,430 \times 10^{-6}$ kg = 1.0134 kg.

Soda requirement for 50,000 L water

$= 106/100[CaSO_4 + MgSO_4 + MgCl_2$ as $CaCO_3$ eq]

× Volume of water

$= 106/100[10.0 + 10.0 + 2.11]$ mg/L × 50,000 L

$= 106/100[22.11]$ mg/L × 50,000 L = 1,171,830 mg

$= 1,171,830 \times 10^{-6}$ kg = 1.171830 kg.

Problem 4. *A water sample on analysis gave the following data:*

$Ca^{+2} = 30$ *mg/L;* $Mg^{+2} = 24$ *mg/L;* $CO_2 = 24$ *mg/L; HCl = 50 mg/L;* $K^+ = 10$ *mg/L.*

Calculate the quantities of Lime (90% pure) and Soda (94% pure) required to soften million litres of water sample.

Solution. Calculation of $CaCO_3$ equivalents:

Costituent	*Multiplication factor*	*$CaCO_3$ equivalent*
$Ca^{+2} = 30$ mg/L	100/40	30 × 100/40 = 75 mg/L
$Mg^{+2} = 24$ mg/L	100/24	24 × 100/24 = 100 mg/L
$CO_2 = 24$ mg/L	100/44	24 × 100/44 = 54.5 mg/L
HCl = 50 mg/L	100/73	50 × 100/73 = 68.5 mg/L

Lime requirement for million litres (10^6 L) of water

$= 74/100[Mg^2 + CO_2 +$ HCl as $CaCO_3$ eq]

× Volume of water (100% purity)

$= 74/100[100 + 54.5 + 68.5]$ mg/L × 10^6 L × (100/90)

$= 74/100[223]$ mg/L × 10^6 L × (100/90)

$= (74 \times 223 \times 10^6)/90$ mg

$= 1.834 \times 10^8$ mg $= 1.834 \times 10^2$ kg = 183.4 kg.

Soda requirement for million litres (10^6 L) of water

$= 106/100[Ca^{+2} + Mg^{+2} +$ HCl as $CaCO_3$ eq]

× Volume of water × (100% purity)

$= 106/100[75 + 100 + 68.5]$ mg/L × 10^6 L × (100/94)

$= 74/100[243.5]$ mg/L × 10^6 L × (100/94)

$= (106 \times 243.5 \times 10^6)/94$ mg

$= 2.946 \times 10^8$ mg $= 2.946 \times 10^2$ kg = 294.6 kg.

EXERCISE

1. What is hard water? Mention the types of hardness and how the hardness is removed from water?
2. How determine the hardness of water by EDTA method?
3. What is potable water? What are the requirements for drinking water?
4. Sketch neat flow diagram of municipal water supply.
5. Write a note on Sterilization and Disinfection.
6. What are Boiler troubles? Explain about the following:
   (*a*) Priming and Foaming
   (*b*) Scale and Sludge formation
   (*c*) Boiler corrosion
   (*d*) Caustic embrittlement.
7. What is softening of water? How the water is softened by the following process?
   (*a*) Lime-Soda process
   (*b*) Zeolite process
   (*c*) Ion exchange process.
8. What is meant by Desalination? How the water is purified by the following process?
   (*a*) Osmosis
   (*b*) Reverse osmosis.
9. What is semi-permeable membrane? What type of membranes are used in desalination?

Chapter **2** ELECTROCHEMISTRY

2.1 INTRODUCTION

Electrochemistry is the branch of chemistry which deals with the transformation of electrical energy into chemical energy and vice versa. It is the study of phenomena at electrode solution interfaces. Electrochemistry deals with the relationship between electrical, chemical phenomena and the laws of interaction of these phenomena. The branch of electrochemistry is of major technical importance.

The passage of electricity through a substance is called electrical conductance. Electrical conductance involves movement of electrons or ions. A substance which allows electric current to pass through it, is called a conductor. Example, all metals fused salts, acids, alkalis.

The electrical conductors are of two types:

1. Metallic or Electronic conductors.
2. Electrolytic conductors.

1. Metallic or Electronic Conductors. Metallic conductors conduct electricity due to the movement of electrons from one end to another end. In a solid, the electrical conduction involves the free movement of electrons in the metal lattice, without any movement of the lattice atom; this type of conduction is called metallic conduction. In metallic conductors, the electricity is carried by the electrons, the atomic nuclei remaining stationary. These conductors are further sub classified into three types.

(*i*) Good conductors

(*ii*) Semi-conductors

(*iii*) Non-conductors or Insulators.

(*i*) **Good conductors.** It is a substance, which conduct electricity fully and freely.

Example. Metals like Cu, Al and Fe.

(*ii*) **Semi-conductors.** A substance, which partially conducts electricity.

Example. Silicon, Germanium.

(*iii*) **Non-conductors or insulators.** It is a substance, which does not conduct electric current *i.e.*, which does not allow the passage of current through it.

Example. Wood.

2. Electrolytic Conductors. It is a substance, its inaqueous solution or in molten state liberates ions and allows electric current to pass through. Electrolytic conductors are further sub classified into three types depending upon the extent of dissociation at ordinary dilutions.

- **Strong electrolytes.** These are completely dissociated into ions at all concentrations.

  Example. NaCl, HCl, NaOH, etc.

- **Weak electrolytes.** Weak electrolytes dissociate only to a small extent even at very high dilutions.

  Example. CH_3COOH, NH_4OH, etc.

- **Non-electrolytes.** Non-electrolytes do not dissociate into ions even at low dilutions.

  Example. Glucose, Sugar.

Differences between metallic and electrolytic conductors.

S.No.	*Metallic conductors*	*Electrolytic conductors*
1.	It involves the flow of electrons in a conductor.	It involves the movement of ions in a solution.
2.	It does not involve any transfer of matter.	It involves transfer of electrolyte in the form of ions.
3.	Generally metallic conduction shows an increase in resistance as the temperature is raised leading to decrease in conduction.	With increase in temperature, the degree of dissociation of electrolyte increases, hence conductivity increases.
4.	No net chemical change takes place.	Chemical reaction takes place at two electrodes.

Conductivity of electrolytes. Electrolytic conduction involves the transfer of electrons through the migration of positive and negative ions towards the electrodes. In an ionic solution, the cations and the anions are free to move and both can transport charge. When a current is passed through the solution, the ions carry a current. The ability of the ions in solution to carry current is conductivity. The conductivity of a solution depends on the number of anions and cations present in it and also on how readily these ions can move. Like electronic conductors, electrolytes also obey Ohm's law.

Resistance (R). The tendency of a material to stop the flow of current through it is known as resistance. It is measured in ohms (Ω).

Ohm's law. The resistance of a conductor is directly proportional to its length (l) and inversely proportional to its cross-sectional area (A).

$$R = \rho \frac{l}{A}$$

where, R = resistance in ohms.

ρ = specific resistance or resistivity.

l = length in cm.

A = area of cross-section in cm^2.

Specific resistance. If l = 1 cm, A = 1 cm^2; R= ρ *i.e.*, specific resistance of a conductor is the resistance between two opposite faces of a centimetre cube of that substance.

Units. Ω cm.

Conductance. Reciprocal of the resistance is known as conductance (C).

$$C = \frac{1}{R}$$

Units. Ω^{-1}.

Specific conductance. Reciprocal of the specific resistance of an electrolyte solution is known as specific conductance (κ).

$$\kappa = \frac{1}{\rho} = \frac{1}{AR}$$

Units. $\Omega^{-1}\ cm^{-1}$.

Equivalent conductance. The conductance of all the ions present in 1 g equivalent of the electrolyte in the solution at given dilution. If 1 g equivalent of electrolyte is contained in ν ml, then;

$$\lambda_{eq} = \nu \times \kappa$$

If normality of electrolytic solution is N, then;

$$\nu = \left(\frac{1}{N}\right) L = \frac{1000}{N}\ ml$$

$$\lambda_{eq} = \frac{1000\ \kappa}{N}$$

Units. $\Omega^{-1}\ cm^2\ eq^{-1}$.

Molar conductivity. The conductance of all the ions present in 1 mole of the electrolyte in the solution is known as molar conductivity (λ_m). If M is the molar conductance in mol/L, then

$$\lambda_m = \frac{1000\ \kappa}{M}$$

Units. $\Omega^{-1}\ cm^2\ mol^{-1}$.

2.2 CONCEPT OF IONIC MOBILITY

At infinite dilution all the electrolytes are completely dissociated, their molar conductances differ vastly from one another. This has been attributed to differences in the speeds of the ions. The speed of an ion varies with the potential applied, it is better to use the term ionic mobility.

The distance travelled by an ion per second under a potential gradient of 1 volt per metre is known as ionic mobility.

Potential gradient is given by the potential difference applied at the electrodes divided by the distance between the electrodes.

Ionic mobility possesses a fixed value for each ion, no matter of which electrolyte it forms a part. It may be emphasized that ionic conductance is not the same thing as ionic mobility. These are however, proportional to each other. If u_+^0 and u_-^0 are the actual velocities of positive and negative ions of a given electrolyte at infinite dilution, then

$$\lambda_+^0 = k\,u_+^0$$

$$\lambda_-^0 = k\,u_-^0$$

where k is a constant.

Absolute ionic velocity of an ion represents its velocity in cm/s under a potential gradient of 1 volt/cm. It has been shown that under a potential gradient of 1 volt/cm, the value of constant 'k' is the change in one gram equivalent of the ion *i.e.*, 96500 coulombs (or 1 Faraday). Mathematically,

$$\text{Ionic mobility} = \frac{\text{Ionic conductance}}{96500}.$$

2.2.1 Arrhenius Ionic Theory

The main postulates of the theory are:

1. In the solutions, all the electrolytes are spontaneously dissociated, to some extent, into charged particles are called ions. The ions carrying positive charge are called cations, while those carrying negative charge are called anions.

   The process of spontaneous splitting up of an electrolyte into ions, when dissolved in a solvent is called ionization (or) electrolytic dissociation.
2. The total positive charge on the cations present in a solution is equal, but opposite to the total negative charge present on the anions. Thus, the solution of an electrolyte is neutral as a whole.
3. There is a state of dynamic equilibrium exists between the ionized and unionized molecules, *i.e.*,

   $$\underset{\text{(Unionized molecules)}}{AB} \rightleftharpoons \underset{\text{(Cation)}}{A^+} + \underset{\text{(Anion)}}{B^-}$$

   Thus, the process of electrolytic dissociation is a reversible process.
4. The ions are free to move and under the influence of electricity, they are directed towards oppositely charged electrodes, *i.e.*, cations move towards cathode; while anions move towards anode. Electric current does not produce ions, but it only has a directive effect on the ions already present in solution.
5. The properties of electrolytes in solution are the properties of ions produced.
6. The electrolyte at a given dilution may not be completely ionized and the fraction of the total molecules ionized, is termed degree of ionization, *i.e.*,

   $$\text{Degree of ionization} = \frac{\text{No. of molecules dissociated into ions}}{\text{Total no. of molecules taken}}.$$

7. The degree of ionization depends upon the following factors:
   (*a*) **Nature of electrolyte.** Some solutes (HCl, NaCl, NaOH, etc.) are highly ionized (strong electrolytes); while the other (NH_4OH, CH_3COOH, etc.) are poorly ionized (weak electrolytes).
   (*b*) **Nature of solvent.** High dielectric solvents like water are highly ionizing, while low dielectric solvents (C_6H_6) are poorly ionized.
   (*c*) **Presence of other solutes.** Presence of other ions also affects the degree of ionization of a solute.Thus, the presence of same ions from another solute depresses the degree of ionization of the given solute.
   (*d*) **Dilution.** Degree of ionization increases with dilution and it becomes unity (maximum) at very high dilution, called infinite dilution.
   (*e*) **Temperature.** The degree of ionization increases with the increase of temperature.
8. The ions behaves like molecules in producing properties like osmotic pressure, depression in freezing point, elevation in boiling point and lowering of vapour pressure.

2.3 KOHLRAUSCH'S LAW

The molar conductivity at infinite dilution for any electrolyte is the sum of molar conductivities for the cation and anion. This is known as Kohlrausch's law.

Mathematically, Kohlrausch's law can be represented as

$$\lambda_m^\infty = \nu_+ \lambda_+^\infty + \nu_- \lambda_-^\infty$$

where ν_+, ν_- are number of anion and cations, and λ_+^∞, λ_-^∞ are the molar conductivities of the cation and anion respectively at infinite dilution.

2.4 APPLICATIONS OF KOHLRAUSCH'S LAW

1. Determination of Molar Conductivity of Weak Electrolytes at Infinite Dilution. Molar conductivity at infinite dilution can be calculated by graphical method. But the graphical method is not applicable to weak electrolytes as they are very slightly ionized. Kohlrausch's law can be used to determine the value of λ_m^∞ for weak electrolytes.

For example, molar conductivity of acetic acid (CH_3COOH) at infinite dilution can be calculated from the molar conductances at infinite dilution of HCl, CH_3COONa, NaCl (all are strong electrolytes) as fallows;

$$\lambda_m^\infty \text{ HCl} - \lambda^\infty \text{ H}^+ + \lambda^\infty \text{ Cl}^-$$

$$\lambda_m^\infty \text{ CH}_3\text{COONa} = \lambda^\infty \text{ Na}^+ + \lambda^\infty \text{ CH}_3\text{COO}^-$$

$$\lambda_m^\infty \text{ NaCl} = \lambda^\infty \text{ Na}^+ + \lambda^\infty \text{ Cl}^-$$

$$\lambda_m^\infty \text{ HCl} + \lambda_m^\infty \text{ CH}_3\text{COONa} - \lambda_m^\infty \text{ NaCl}$$

$$= [\lambda^\infty \text{ H}^+ + \lambda^\infty \text{ Cl}^-] + [\lambda^\infty \text{ Na}^+ + \lambda^\infty \text{ CH}_3\text{COO}^-] - [\lambda^\infty \text{ Na}^+ + \lambda^\infty \text{ Cl}^-]$$

$$= [\lambda^\infty \text{ H}^+ + \lambda^\infty \text{ CH}_3\text{COO}^-] = \lambda_m^\infty \text{ CH}_3\text{COOH}$$

2. Determination of Degree of Dissociation. The term degree of dissociation is the fraction of the total number of molecules dissociated into ions, *i.e.,*

$$\alpha_c = \frac{\text{No. of molecules dissociated into ions}}{\text{Total number of molecules taken}}$$

We also know that the conductivity of a solution is due to the presence of ions in solution. Greater the number of ions greater is the conductivity.

$\therefore$ λ_m^c, the molar conductivity at a particular dilution α No. of molecules dissociated into ions at this dilution. ...(*i*)

But at infinite dilution all molecules are in ionic form,

$\therefore$ λ_m^∞ α Total number of molecules taken. ...(*ii*)

Dividing (*i*) by (*ii*), we get:

$$\frac{\lambda_m^c}{\lambda_m^\infty} = \frac{\text{Number of molecules dissociated into ions at a particular dilution}}{\text{Total number of molecules taken}} \quad ...(iii)$$

$$\frac{\lambda_m^c}{\lambda_m^\infty} = \text{Degree of ionization} = \alpha$$

Thus, degree of dissociation at any dilution is the ratio of the molar conductivity at that dilution to the molar conductivity at infinite dilution.

The value of λ_m^c can be obtained by direct measurement; while λ_m^∞ can be obtained with the help of Kohlrausch's law, *i.e.,*

$$\lambda_m^\infty = \nu_+ \lambda_+^\infty + \nu_- \lambda_-^\infty$$

Hence, the value of α_c can be calculated.

3. Determination of Solubility of Sparingly Soluble Salts. For ordinary purposes, ionic salts like AgCl, $BaSO_4$, $CaCO_3$, $PbSO_4$, PbS, $Fe(OH)_3$, etc., are regarded as insoluble, but they do have a very small, but definite solubilities in water. The solubility of such sparingly soluble salt is obtained by determining the specific conductivity (κ) of a saturated salt solution. Since only a very small amount of salt is present in solution, the equivalent conductivity at such high dilution can particularly be taken as λ_{eq}^∞.

For very sparingly soluble salts $\lambda_{eq}^c \cong \lambda_{eq}^\infty = \kappa\nu^-$

The value of solubility can be calculated with the help of Kohlrausch's law.

$$\nu = \frac{\lambda_{eq}^c}{\kappa} = \frac{\lambda_{eq}^\infty}{\kappa} = \frac{\lambda_+^\infty + \lambda_-^\infty}{\kappa}$$

But ν cm³ of saturated solution contains = 1 g equivalent

1000 cm³ or 1 L of saturated solution contains = $\left(\frac{1000}{\nu}\right)$ g equivalent

Hence, solubility of salt, $S = \left(\frac{1000}{\nu}\right)$ g eq/L $= \frac{1000\ E}{\nu}$ g/L $= \frac{1000 \times E \times \kappa}{[\lambda_+^\infty + \lambda_-^\infty]}$ g/L

where E is the equivalent weight of salt.

2.5 CONDUCTOMETRIC TITRATIONS

Those titrations in which the end point is determined by using conductivity measurements are known as conductometric titrations. This titration method is based on the fact that conductance of an aqueous solution, containing an electrolyte, depends upon:

(*i*) The number of free ions in the solution,

(*ii*) The charge on the free ions, and

(*iii*) Mobility (or speed) of the ions.

During course of titration the number of free ions in the solution changes. Not only that, even the identity of the ions also changes. As a result of this, conductance of the solution also undergoes a change. Let us take few cases to illustrate.

1. Acid-Base titrations
2. Precipitation titrations.

1. Acid-Base Titrations

(*i*) Titration of strong acid (like HCl) with strong base (like NaOH). The initial conductivity of the acid solution is high due to the presence of highly mobile H^+ ions by the addition of base from the burette to the acid solution present in conductivity bottle, conductivity value is decreased due to the replacement of highly mobile H^+ ions with slow moving Na^+ ions till the H^+ ions are completed. On further addition of NaOH to the acid solution the conductivity will rise due to the presence of only OH^- ions. At the minimum point, there is no excess of either acid or base and hence, it corresponds to the end point.

$$\underset{\text{Strong acid}}{[H^+ Cl^-]} + \underset{\text{Strong base}}{[Na^+ OH^-]} \longrightarrow \underset{\text{Salt}}{Na^+Cl^-} + \underset{\text{Unionized}}{H_2O}$$

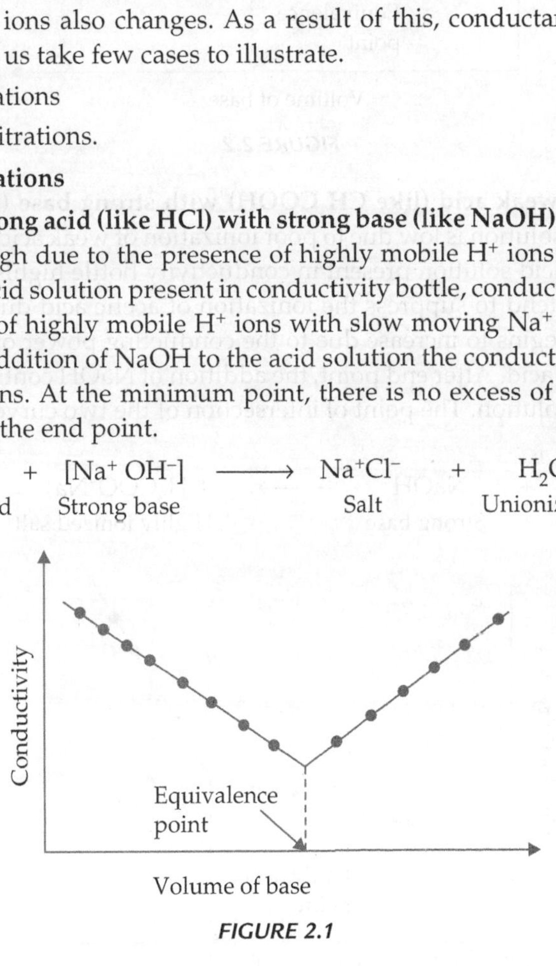

FIGURE 2.1

(*ii*) Titration of strong acid (like HCl) with weak base (like NH_4OH). The initial conductivity of the acid solution is high due to the presence of highly mobile H^+ ions by the addition of base from the burette to the acid solution present in conductivity bottle conductivity value is decreased due to the replacement of highly mobile H^+ ions with slow moving NH_4^- ions till the H^+ ions are completed. On further addition of NH_4OH to the acid solution does not cause any appreciable change in conductance. At the minimum point, there is no excess of either acid or base and hence, it corresponds to the end point.

$$\underset{\text{Strong acid}}{H^+Cl^-} + \underset{\text{Weak base}}{NH_4OH} \longrightarrow \underset{\text{Salt}}{NH_4^+\,Cl^-} + \underset{\text{Unionized}}{H_2O}$$

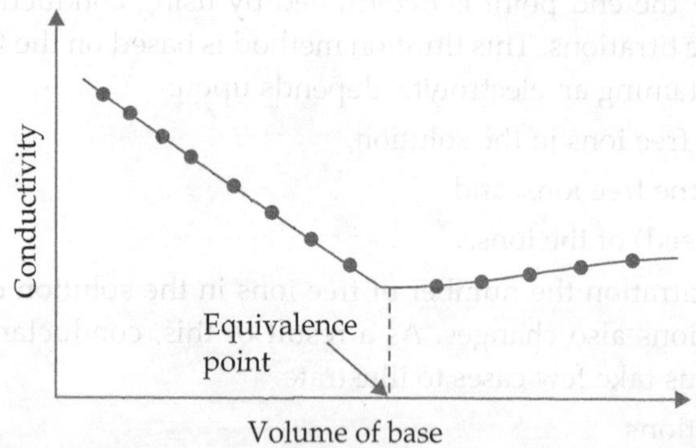

FIGURE 2.2

***(iii)* Titration of weak acid (like CH_3COOH) with strong base (like NaOH).** The initial conductivity of the acid solution is low due to poor ionization of weak acid. By the addition of base from the burette to the acid solution present in conductivity bottle highly ionized salt is formed. The acetate ions at first tend to suppress the ionization of acetic acid due to common ion effect. Later the conductivity begins to increase due to the conducting power of the highly ionized salt exceeds that of the weak acid. After end point, the addition of NaOH contributes sharp increase in the conductivity of the solution. The point of intersection of the two curves gives the end point of the titration.

$$\underset{\text{Weak acid}}{CH_3COOH} + \underset{\text{Strong base}}{NaOH} \longrightarrow \underset{\text{Highly ionized salt}}{CH_3COO^-Na^+} + \underset{\text{Unionized}}{H_2O}$$

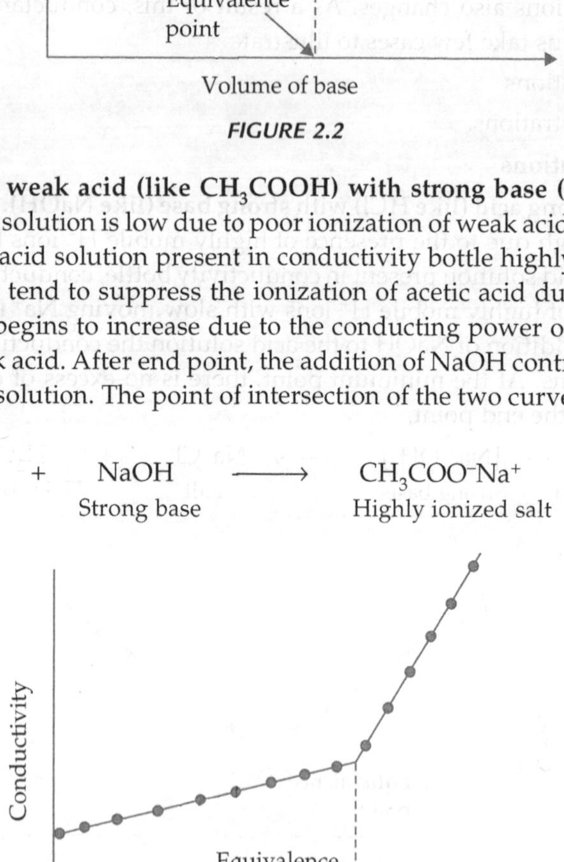

FIGURE 2.3

***(iv)* Titration of weak acid (like CH_3COOH) with weak base (like NH_4OH).** The addition of NH_4OH to the CH_3COOH solution may even cause a decrease in conductance in the beginning, because the common ions formed depress the dissociation of their respective electrolyte (*i.e.*, CH_3COOH and NH_4OH).

However, on further addition of, an increase in the conductivity of the solution results, because the conductance of the highly ionized salt (*e.g.,*) exceeds the conductance of the weak acid(CH_3COOH) it replaces. After the neutralization of the acid, further addition of poorly ionized does not cause any appreciable change in the conductance shown in Fig. 2.4.

$$\underset{\text{Weak acid}}{CH_3COOH} + \underset{\text{Weak base}}{NH_4OH} \longrightarrow \underset{\text{Highly ionized salt}}{NH_4^+} + CH_3COO^- + H_2O$$

$$NH_4^+ + CH_3COO^- \xrightarrow{NH_4OH} \text{No change in conductance}$$

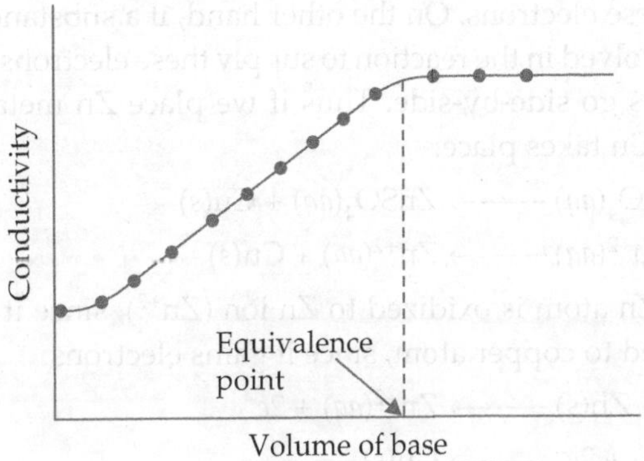

FIGURE 2.4

2. Precipitation Titrations. Precipitation titrations can also easily be followed conductometrically.

Example. Titration of KCl solution with $AgNO_3$ solution.

The titration of KCl solution against $AgNO_3$ solution, the reaction proceeds as follows;

$$[K^+ + Cl^-] + [Ag^+ + NO_3^-] \longrightarrow AgCl\downarrow + [K^+ + NO_3^-]$$

Here AgCl gets precipitated; while the K^+ ions remain unchanged. As a result of this, Cl^- ions in solution are now replaced by free NO_3^- ions. The mobility of Cl^- and NO_3^- ions are practically the same, so the conductance of the solution with the addition of $AgNO_3$ will not change. When the equivalence point is reached, further addition of $AgNO_3$(addition of free Ag^+ and NO_3^-) to the solution brings a steep rise in the conductance of the solution. The sharp change of the curve is shown in Fig. 2.5.

$$AgCl\downarrow + [K^+ + NO_3^-] \xrightarrow{AgNO_3} \text{Conductance increases}$$

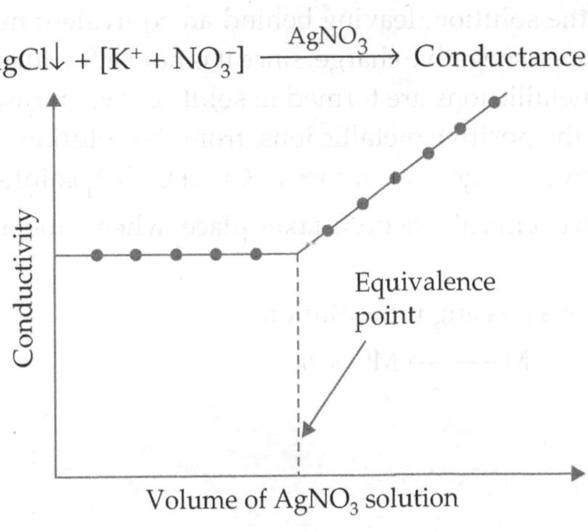

FIGURE 2.5

Redox Reactions. "The reactions in which both oxidation and reduction reactions takes place are known as ***redox reactions***".

Oxidation. Oxidation is a process which involves loss of electrons by a substance.

$$Zn(s) \longrightarrow Zn^{+2}(aq) + 2e^- \qquad \text{(oxidation)}$$

Reduction. Reduction is a process which involves gain of electrons by a substance.

$$Cu^{+2}(aq) + 2e^- \longrightarrow Cu(s) \qquad \text{(reduction)}$$

It is quite apparent that if a substance loses electrons, some other substance must be involved in the reaction to accept these electrons. On the other hand, if a substance accepts electrons, some other substance must be involved in the reaction to supply these electrons. In other words, oxidation and reduction must always go side-by-side. Thus if we place Zn metal in a solution of $CuSO_4$, immediate preparation of Cu takes place.

$$Zn(s) + CuSO_4(aq) \longrightarrow ZnSO_4(aq) + Cu(s)$$

$$Zn(s) + Cu^{+2}(aq) \longrightarrow Zn^{+2}(aq) + Cu(s)$$

In this change, the Zn atom is oxidized to Zn ion (Zn^{+2}), since it loses electrons, while the copper ion (Cu^{+2}) is reduced to copper atom, since it gains electrons.

(*i*) $$Zn(s) \longrightarrow Zn^{+2}(aq) + 2e^- \qquad \text{(oxidation)}$$

(*ii*) $$Cu^{+2}(aq) + 2e^- \longrightarrow Cu(s) \qquad \text{(reduction)}$$

$$Zn(s) + Cu^{+2}(aq) \longrightarrow Zn^{+2}(aq) + Cu(s) \qquad \text{(redox)}$$

The overall reaction obtained by adding the (*i*) and (*ii*) is called ***redox*** or ***oxidation reduction reaction.***

2.6 ELECTRODE POTENTIAL

"The tendency of an electrode to lose or gain electrons, when it is in contact with its own ions is known as ***Electrode potential.***"

A metal (M) consists of metal ions (M^{n+}), with the valency electrons that bind them together. Now, if a metal is in contact with a solution of its own salt, the positive ions in the metal come into equilibrium with those in the solution, leaving behind an equivalent number of electrons on the metal. Thus, the metal acquires a negative charge, since it is now left with excess number of electrons and a number of positive metallic ions are formed in solution *e.g.*, in case of Zn in $ZnSO_4$ solution [Fig. 2.6(*a*)]. Conversely, if the positive metallic ions, from the solution, enter the metallic lattices, then metal acquires a positive charge, *e.g.*, in case of Cu in $CuSO_4$ solution [Fig. 2.6(*b*)].

Thus, following two chemical reactions takes place, when a metal is in contact with its salt solution.

1. Positive metallic ions passing into solution

$$M \longrightarrow M^{n+} + ne^- \qquad \text{(oxidation)}$$

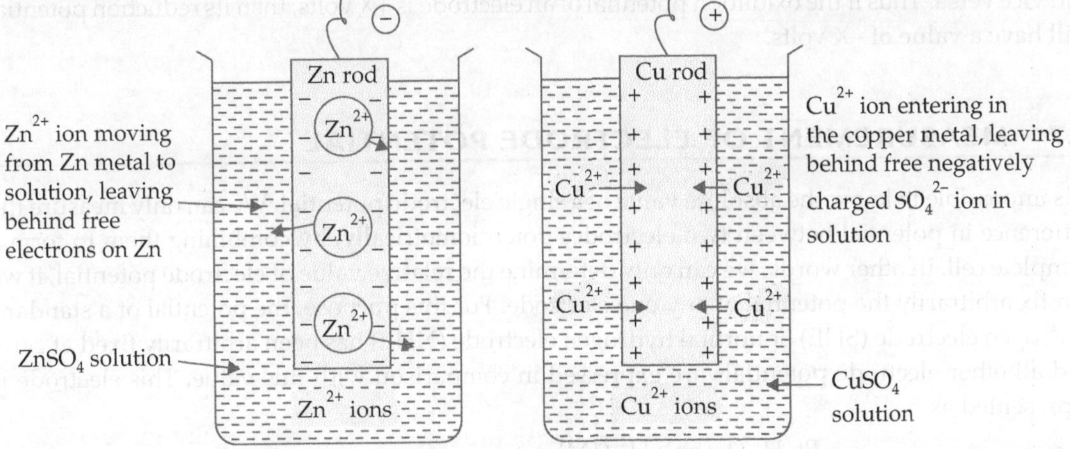

(*a*) De-electronation (*b*) Electronation

***FIGURE 2.6* Electrode potential.**

When '*n*' electrons are left behind on the metal, and it acquires a negative charge. The rate of this reaction depends on (*i*) the nature of the metal, (*ii*) the temperature, and (*iii*) the concentration of metal ions in solution.

2. Positive ions depositing on the metal electrode

$$M^{n+} + ne^- \longrightarrow M \qquad \text{(reduction)}$$

When metal acquires a positive charge. The rate of this reaction depends on the above three factors.

When a metal is placed in the solution of its own salt, the chemical reaction (1) and (2), as the case may be, takes place and ultimately a dynamic equilibrium is established, because –ve or +ve charge developed on the metal attracts the positively or negatively charged free ions in the solution. Due to this attraction, the positive or negative ions remain quite close to the metal. Thus, a short of layer of +ve ions in Fig. 2.6(*a*) or –ve ions in Fig. 2.6(*b*) is formed all around the metal. This layer is called ***Helm holtz Electrical double layer***. A difference of potential is, consequently set up between the metal and the solution. This potential difference will persist as long as the charge is allowed to remain on the metal, and this will prevent any further passing of the +ve ions from or to the metal.

At equilibrium, the potential difference between the metal and the solution becomes a constant value. The equilibrium, potential difference so established is called the *"**electrode potential**"* of the metal. Thus,

- The measure of tendency of a metallic electrode to lose or gain of electrons, when it is in contact with a solution of its own salt of unit molar concentration at 25°C.
- Consequently, the tendency of an electrode to lose electrons is a direct measure of its tendency to get oxidized and this tendency is called *"**Oxidation potential**"*.
- Similarly, the tendency of an electrode to gain electrons is a direct measure of its tendency to get reduced, and this tendency is known as *"**Reduction potential**"*.

It is quite obvious that the value of reduction potential is negative of its oxidation potential and vice versa. Thus if the oxidation potential of an electrode is +X volts, then its reduction potential will have a value of –X volts.

2.7 MEASUREMENT OF ELECTRODE POTENTIAL

It is impossible to know the absolute value of a single electrode potential. We can only measure the difference in potential between two electrodes potentiometrically, by combining them to form a complete cell. In other words, we can only determine the relative value of electrode potential, if we can fix arbitrarily the potential of any one electrode. For this purpose, the potential of a standard hydrogen electrode (SHE) or normal hydrogen electrode (NHE) has been arbitrarily fixed at zero, and all other electrode potentials are expressed in comparison with this value. This electrode is represented as,

$$Pt, H_2 \text{ (1 atm)}; H^+ \text{ (1M)}$$

and either of the following reactions may take place at this electrode

$$H^+(aq) + e^- \longrightarrow \frac{1}{2} H_2(g)$$

$$\frac{1}{2} H_2(g) \longrightarrow H^+(aq) + e^-$$

Through a salt bridge. Thus in order to measure the electrode potential, we couple this electrode with SHE through a salt bridge. The voltmeter reading directly gives the electrode potential, since the potential of SHE is zero.

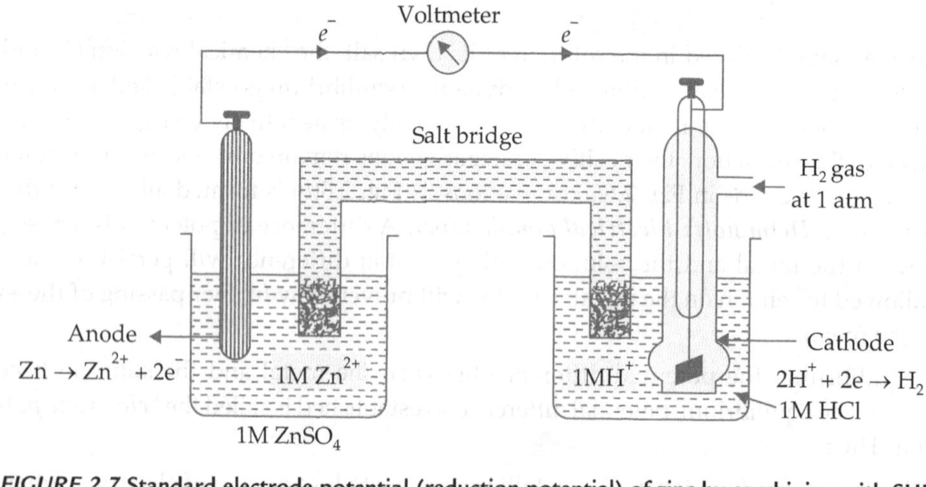

FIGURE 2.7 **Standard electrode potential (reduction potential) of zinc by combining with SHE.**

As zinc electrode acts as anode in this case, the electrode potential 0.76 V, is oxidation potential. Consequently, reduction potential of Zn electrode is –0.76 V. According to modern conventions, if on coupling of an electrode with SHE, reduction occurs at the given electrode, the electrode potential (reduction) is given a positive sign (+ve), and if oxidation occurs at the given electrode, the electrode potential (reduction) is given a negative sign (–ve).

Importance of an Electrode Potential. When elements are arranged in increasing order of their standard electrode potential in a series is called electrochemical series. The electrochemical series provide valuable informations.

1. Relative ease of oxidation or reduction. A system with high reduction potential has a great tendency to undergo reduction, so we can easily interpret the behaviour of different elements. For example, the standard reduction potential of F_2/F system is highest (*i.e.*, +2.87 volts), so F ions are very easily reduced to F_2 or conversely, F_2 is oxidised with great difficulty to F ions. On the other hand, standard reduction potential of Li^+/Li system is least (*i.e.*, –3.05 volts), so Li^+ is reduced with great difficulty to Li or conversely, Li is very easily oxidised to Li^+.

2. Replacement tendency. Higher value of reduction potential shows a great tendency to assume the reduced form, so knowledge of electrode potential gives an indication of relative replacement tendency. For example, we can know whether Cu will displace Zn from solution or vice versa. We know that standard electrode potentials,

$$E_{Cu+2/Cu} = 0.34 \text{ V} \quad \text{and} \quad E_{Zn+2/Zn} = -0.76 \text{ V}$$

So, Cu^{+2} has a greater tendency to acquire Cu form than Zn^{+2} has for acquiring Zn form. In other words, zinc will displace copper from the solution of latter, or the reaction will occur in the direction.

$$Zn + Cu^{+2} \longrightarrow Zn^{+2} + Cu$$

3. Predicting spontaneity of redox reactions. Spontaneity of a redox reaction can be predicted from the e.m.f. (E) value of the complete cell reaction. Positive value of E of a cell reaction indicates that the reaction is spontaneous. It value of E is negative, the reaction is not feasible. In general, an element having lower reduction potential can displace another metal having higher reduction potential from its salt solution spontaneously.

4. Calculation of equilibrium constant. The standard electrode potential

$$E^0 = \frac{RT}{nF} \ln K_{eq} = \frac{2.303 \text{ RT}}{nF} \log K_{eq}$$

$$\log K_{eq} = \frac{nF \times E^0}{2.303 \text{ RT}} = \frac{nE^0}{0.0592 \text{ V}} \quad \text{(at 25°C)}$$

From the value of standard electrode potential (E^0) for a cell reaction, its equilibrium constant (K_{eq}) can be calculated.

2.8 SINGLE ELECTRODE POTENTIAL

The metal rod dipped in the aqueous solution of its salt or a gaseous non-metal in contact with a solution containing its anion in the form of salt (or acid or alkali) functions as single electrode. The metal electrodes like Zn electrode $Zn^{+2}(aq)/Zn$, copper electrode $Cu^{+2}(aq)/Cu$ etc., and non-metal electrodes like Hydrogen electrode, $H^+(aq)/H_2(g)$, Pt, Chlorine electrode, Pt; $Cl_2(g)/Cl(aq)^-$, Pt etc.

These single electrode, exhibits its characteristic electrode potential. This potential is called ***"single electrode potential"*** (E). The value of this potential depends on the chemical identity of the metal or non metal, the concentration of the concerned ion in the aqueous solution and temperature.

The potential exhibited by a single electrode at unit concentration of the concerned metal ion or non-metal ion solution at 25°C is called *"**standard single electrode potential**"* (E^0).

2.9 NERNST EQUATION

The Nernst equation is

$$E = E^0 + \frac{0.0592}{n}\log[M^{n+}] \quad \text{or} \quad E = E^0 + \frac{2.303\ RT}{nF}\log[M^{n+}]$$

E – Electrode potential R – Gas constant

E^0 – Standard electrode potential T – Temperature

n – Number of valency electrons F – Faraday (96,500 coulombs)

Derivation of Nernst equation: Consider a general redox reaction

$$M^{n+}(aq) + ne^- \rightleftharpoons M(s)$$

Now for a reversible reaction, the free energy change (ΔG) and its equilibrium constant (K) are inter-related as

$$\Delta G = RT \ln K + RT \ln\frac{[\text{product}]}{[\text{reactant}]}$$

$$\Delta G = \Delta G^0 + RT \ln\frac{[\text{product}]}{[\text{reactant}]} \qquad ...(1)$$

where ΔG^0 is known as the standard free energy change (*i.e.*, the change in free energy, when the concentrations of the reactants and the products are unity each).The above equation, is known popularly as Van't Hoff reaction isotherm. In a reversible reaction, the electrical energy is produced at the expanse of free energy decrease, *i.e.*,

$$\Delta G = nEF \quad \text{and} \quad \Delta G^0 = -nE^0F$$

Substitute ΔG and ΔG^0 values in equation (1)

$$-nEF = -nE^0F + RT \ln \frac{[M]}{[M^{n+}]}$$

$$-nEF = -nE^0F + RT \ln\frac{1}{[M^{n+}]} \qquad \text{[concentration of metal [M] =1]}$$

$$-nEF = -nE^0F - RT \ln [M^{n+}]$$

$$E = E^0 + \frac{RT}{nF}\ln [M^{n+}]$$

$$E = E^0 + \frac{2.303\ RT}{nF}\log [M^{n+}]$$

This is known as Nernst equation of electrode potential at 25°C.

$$\frac{2.303\ RT}{nF} = \frac{0.0592}{n}$$

$$\therefore \quad E = E^0 + \frac{0.0592}{n}\log [M^{n+}]$$

2.10 ELECTROCHEMICAL CELLS OR GALVANIC CELLS

"The device in which redox reaction is utilized to get electrical energy from chemical energy is called *Electrochemical cell*".

Or

"The devices which used to convert the chemical energy produced in a redox reaction into electrical energy are known as *Electrochemical cells*".

An electrochemical cell is also commonly referred to as voltaic or galvanic cell. The electrode where oxidation occurs is called anode, while the electrode where reduction occurs is called cathode.

Example. Daniel cell.

Daniel cell. The practical application of an electrochemical or galvanic cell is Daniel cell. It consists of Zn electrode, dipping in $ZnSO_4$ solution (where oxidation takes place) and a copper electrode, dipping in $CuSO_4$ solution (where reduction takes place). In other words, each electrode may be regarded as a half-cell. The two solutions are separated by a salt bridge. The two solutions can seep through the salt bridge, and so come in contact with each other automatically.

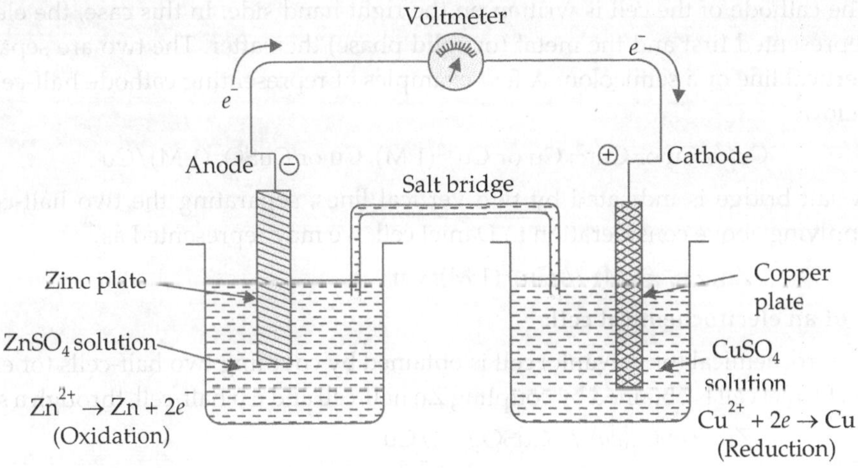

FIGURE 2.8 **Daniel cell.**

The electrode reactions in Daniel cell are

At Anode: $Zn \longrightarrow Zn^{+2} + 2e^-$

At Cathode: $Cu^{+2} + 2e^- \longrightarrow Cu$

Cell reaction: $Zn + Cu^{+2} \longrightarrow Zn^{+2} + Cu$

The tendency of Zn to form Zn^{+2} is greater than the tendency Zn^{+2} to get deposited as Zn and hence, Zn metal acquires a negative charge. On the other hand, tendency of copper to go into solution is less than the tendency of Cu^{+2} to get deposited as Cu and hence, copper electrode becomes positively charged. The e.m.f. of the cell is 1.1 volts.

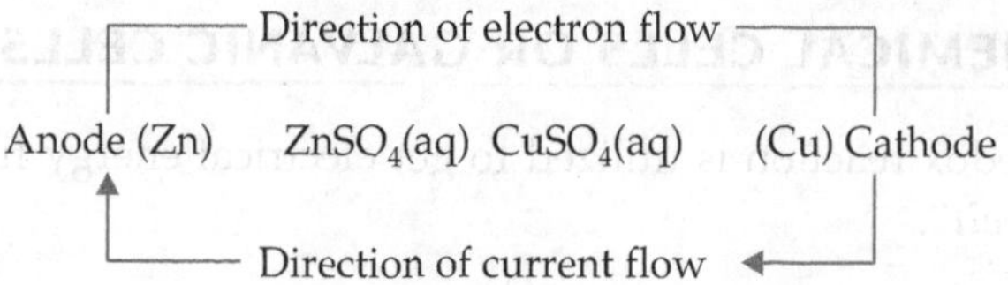

FIGURE 2.9

Representation of a Galvanic cell. According to present conventions, a galvanic cell is represented by keeping in view the following points:

1. Anode is written on the left-hand side, while cathode is written on the right-hand side.
2. The electrode on the left (*i.e.*, Anode) is written by writing the metal (or solid phase) first and then the electrolyte. The two are separated by a vertical line or a semicolon. The electrolyte may be represented by the formula of the whole compound or by ionic species. Additional information regarding concentration may also be mentioned in bracket. These points will be made clear by following examples of representing anode half-cell as,

$$Zn/Zn^{+2} \text{ or } Zn; Zn^{+2} \text{ or } Zn; ZnSO_4 \text{ (1 M)}; Pt, H_2 \text{ (1 atm)}; H^+ \text{ (1M)}$$

3. The cathode of the cell is written on the right-hand side. In this case, the electrolyte is represented first and the metal (or solid phase) thereafter. The two are separated by a vertical line or a semicolon. A few examples of representing cathode half-cell are cited below,

$$Cu^{+2}/Cu \text{ or } Cu^{+2}; Cu \text{ or } Cu^{+2}\text{(1 M)}; Cu \text{ or } CuSO_4 \text{ (1 M)}/Cu$$

4. A salt bridge is indicated by two vertical lines, separating the two half-cells. Thus, applying above consideration to Daniel cell, we may represented as,

$$Zn; Zn^{+2}\text{(1 M)} \;//\; Cu^{+2}\text{(1 M)}; Cu$$

EMF of an electrochemical cell:

An electrochemical (or galvanic) cell is obtained by coupling two half-cells (or electrodes). For example, Daniel cell is obtained by coupling Zn half-cell and Cu half-cell, through a salt bridge.

$$Zn/ZnSO_4(aq) \;//\; CuSO_4(aq)/Cu$$

Mathematically, the e.m.f. of an electrochemical cell is the algebric sum of the single electrode potential, provided proper signs are given according to the actual reaction taking place on the electrodes. Cell is generally written the negative electrode to the left. Based on this, the e.m.f. of galvanic cell can readily be calculated from the reduction half-cell potentials, using the following relationship.

$$E_{cell} = E_{right} - E_{left}$$

where

E_{cell} = E.M.F. of the cell

E_{right} = reduction potential of right-hand side electrode

E_{left} = reduction potential of left-hand side electrode

It should be remember that a positive value of E_{cell} indicates that the cell reaction is feasible. If however, it comes out to be negative, the cell reaction is not feasible and in actual operation of the cell, the electrodes will have to be reversed in order to bring about cell reaction.

Nernst equation for a cell reaction:

$$Zn(s)/Zn^{+2}(aq)//Cu^{+2}(aq)/Cu(s)$$

Reaction is $Zn(s) + Cu^{+2}(aq) \longrightarrow Zn^{+2}(aq) + Cu(s)$

$$\therefore \quad E_{cell} = E^0_{cell} - \frac{2.303\ RT}{nF}\log Q \qquad \left[\because\ Q = \frac{[Zn^{+2}]}{[Cu^{+2}]}\right]$$

At 298 K or 25°C, the Nernst equation can be written as,

$$E_{cell} = [E^0_{cathode} - E^\circ_{anode}] - \frac{0.0592\ V}{n}\log\frac{[Zn^{+2}]}{[Cu^{+2}]}$$

$$= \left[E^0_{Cu^{+2}/Cu} - E^0_{Zn^{+2}/Zn}\right] + \frac{0.0592\ V}{n}\log\frac{[Cu^{+2}]}{[Zn^{+2}]}$$

2.11 ELECTROCHEMICAL SERIES

The electrochemical series is built up by arranging various redox equilibria in order of their standard electrode potentials (redox potentials). The most negative E^0 values are placed at the top of the electrochemical series, and the most positive at the bottom.

The electrochemical series:

Equilibrium			*E^0 (volts)*
$Li^+(aq) + e^-$	$\rightleftharpoons$	$Li^+(s)$	−3.03
$K^+(aq) + e^-$	$\rightleftharpoons$	$K(s)$	−2.92
$Ca^{2+}(aq) + 2e^-$	$\rightleftharpoons$	$Ca(s)$	−2.87
$Na^+(aq) + e^-$	$\rightleftharpoons$	$Na(s)$	−2.71
$Mg^{2+}(aq) + 2e$	$\rightleftharpoons$	$Mg(s)$	−2.37
$Al^{3+}(aq) + 3e^-$	$\rightleftharpoons$	$Al(s)$	−1.66
$Zn^{2+}(aq) + 2e^-$	$\rightleftharpoons$	$Zn(s)$	−0.76
$Fe^{2+}(aq) + 2e^-$	$\rightleftharpoons$	$Fe(s)$	−0.44
$Pb^{2+}(aq) + 2e^-$	$\rightleftharpoons$	$Pb(s)$	−0.13
$2H^+(aq) + 2e^-$	$\rightleftharpoons$	$H_2(g)$	0
$Cu^{2+}(aq) + 2e^-$	$\rightleftharpoons$	$Cu(s)$	+0.34
$Ag^+(aq) + e^-$	$\rightleftharpoons$	$Ag(s)$	+0.80
$Au^{3+}(aq) + 3e^-$	$\rightleftharpoons$	$Au(s)$	+1.50

2.12 POTENTIOMETRIC TITRATIONS

A titration in which the end point is determined with the help of the measurement of the potentials of the reaction mixture is known as potentiometric titration.

In this method, the changes in the electrode potential upon the addition of the titrant are noted against the volume of the titrant added. At the end point the rate of change of the potential is maximum. The end point is determined by plotting a curve of potential versus volume of the titrant and recording the inflexion point, a point where the slope of the curve is maximum as shown in Fig. 2.10(*a*). Alternatively, change in E.M.F. with every small addition of titrant (*i.e.*, $\Delta E/\Delta V$) is plotted against volume V shown in Fig. 2.10(*b*).

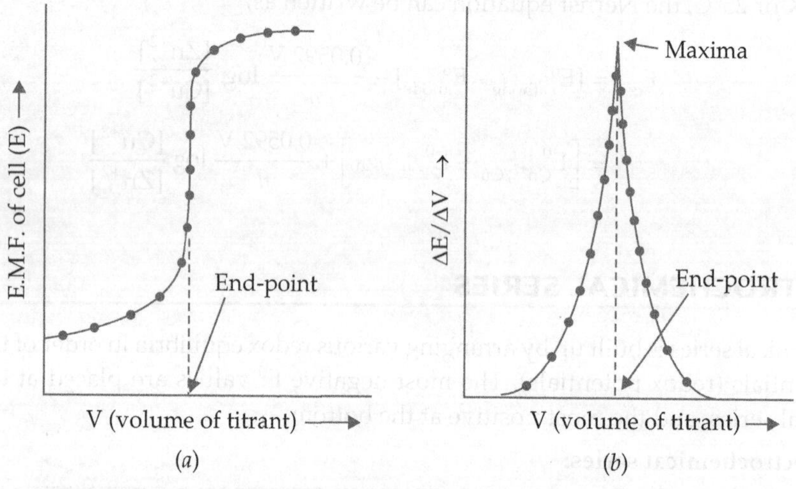

FIGURE 2.10

There are three important types of potentiometric titrations, which are described below:

1. Acid-Base titrations.
2. Oxidation-Reduction titrations.
3. Precipitation titrations.

1. Acid-Base titrations. In acid-base titration quinhydrone electrode is used as indicator electrode. The reference electrode is generally the saturated calomel electrode.

A definite volume of the given acid solution is taken in a beaker. To it a pinch of quinhydrone is added and a stirrer and platinum electrode is dipped in it. This electrode is then connected to a saturated calomel electrode through a potentiometer. On adding standard alkali solution from the burette, the E.M.F of the cell increases at first gradually, but at the end-point the rate of change of potential will be suddenly quite large. The end-point of the titration is then located by plotting $\Delta E/\Delta V$ versus V as shown in Fig. 2.11 and the maximum point on the curve gives the end-point.

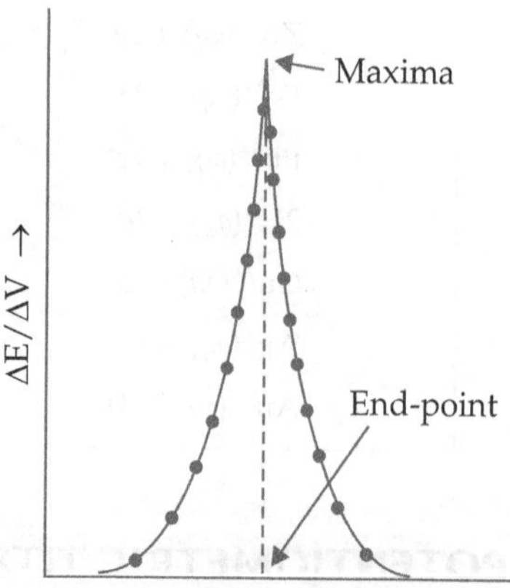

FIGURE 2.11

2. Oxidation-Reduction titrations. Titrations involving oxidizing agents ($K_2Cr_2O_7$ or $KMnO_4$) and reducing agents (like ferrous salt) can be followed potentiometrically by using platinum indicator electrode. On adding $K_2Cr_2O_7$ from the burette to ferrous salt present in the beaker, E.M.F. of the cell will increase first slowly, but at the end-point (equivalence point), there will be sudden jump in potential, since change in the ratio of Fe^{2+}/Fe^{3+} ions concentration, is quite rapid at the equivalence point.

$$E = E^0 + \frac{RT}{F} \ln \frac{(Fe^{2+})}{(Fe^{3+})}$$

3. Precipitation titrations. Titration involving precipitation (*e.g.*, $AgNO_3$ with KCl or KI) can also be followed potentiometrically by the use of active metal electrode. Consider the method of titration of $AgNO_3$ with a standard KCl solution run from the burette. Ag electrode dipping in $AgNO_3$ solution is used as indicator electrode. The potential of Ag^+/Ag half-cell is measured by connecting with saturated calomel electrode through a salt bridge of NH_4NO_3.

As the titration proceeds, the concentration of Ag^+ ions decreases, due to the precipitation of Ag^+ ions as AgCl.

$$Ag^+NO_3^- + K^+Cl^- \longrightarrow AgCl + K^+NO_3^-$$

The reduction potential of indicator electrode (Ag) also goes on decreasing progressively on the addition of KCl. At the end-point, the concentration change of Ag^+ is quite rapid. At the end-point, the Ag^+ concentration becomes very small on account of slight solubility of AgCl. If addition of KCl is continued, the Ag^+ ion concentration is not affected, except a very small decrease, due to decrease in solubility of AgCl on account of common ion (Cl^-) effect. Therefore, addition of KCl beyond end-point causes only a very small decrease in electrode potential or E.M.F. of the cell. On plotting $\Delta E/\Delta V$ versus V, the point of maxima gives the end-point.

Advantages of potentiometric titrations:

1. Coloured solutions, where the use of indicator is impossible are estimated by potentiometric titrations.
2. There is no problem with regard to the choice of indicators based on pH value of the solutions.
3. Polybasic acids can be titrated in steps corresponding to different steps of neutralization.
4. Dependence on colour and external indicators are avoided for redox titration by potentiometric titrations.
5. Solutions containing more than one halide can be analyzed in a single titration against silver nitrate.

2.13 CONCENTRATION CELLS

A cell in which both the electrodes are of the same type but the solutions of electrolyte in which they dip have different concentration is called concentration cell. In concentration cells electrical energy is produced from the transfer of a substance from the solution of a higher concentration to solution of lower concentration.

Concentration cell is made up of two half-cells having identical electrodes, identical electrolyte, except that the concentrations of the reactive ions at the two electrodes are different. The two half-cells may be joined by a salt bridge.

$$- Ag \mid \underset{\text{(Dilute)}}{AgNO_3 (C_1)} \mid \underset{NH_4NO_3}{\text{Salt bridge of satd.}} \mid \underset{\text{[Concentrated]}}{AgNO_3 (C_2)} \mid Ag^+$$

$$- Ag \mid AgNO_3 (C_1M) \mid\mid AgNO_3 (C_2M) \mid Ag^+$$

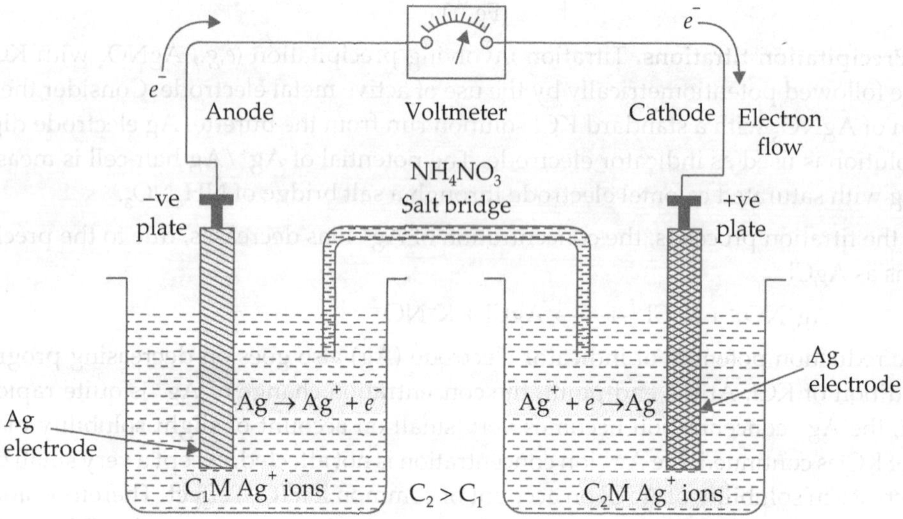

FIGURE 2.12

Theory. When a metal (M) electrode is dipped in a solution containing its own ions [M^{n+}], then a potential is developed at the electrode, the value of which varies with the concentration (C) of the ions in accordance with the Nernst equation.

$$E = E^0 + \frac{2.303\,RT}{nF} \log C$$

Let us consider a general concentration cell represented as,

$$^{-}M \mid M^{n+}(C_1) \mid\mid M^{n+}(C_2) \mid M^+ \text{ (}C_2\text{ is greater than }C_1\text{)}$$

where C_1 and C_2 are the concentrations of active metal ions [M^{n+}] in contact with two electrodes respectively and $C_2 > C_1$.

$$\text{E.M.F. of cell} = E_{right} - E_{left}$$

$$E_{cell} = \left[E^0 + \frac{2.303\,RT}{nF} \log C_2\right] - \left[E^0 + \frac{2.303\,RT}{nF} \log C_1\right]$$

$$E_{cell} = \frac{2.303\,RT}{nF} \log \frac{C_2}{C_1} \quad \text{or} \quad E_{cell} = \frac{0.0592\,V}{n} \log \frac{C_2}{C_1}$$

In this cell, the following reactions occur:

At left electrode (anode): $M \longrightarrow M^{n+}(C_1) + ne^-$

At right electrode (cathode): $M^{n+}(C_2) + ne^- \longrightarrow M$

Cell reaction: $M^{n+}(C_2) \longrightarrow M^{n+}(C_1)$

Evidently, the E.M.F. so-developed is due to the mere transference of metal ions from the solution of higher concentration (C_2) to the solution of lower concentration (C_1).

There are two types of concentration cells,

1. Electrode concentration cells.
2. Electrolyte concentration cells.

1. Electrode concentration cells. In these cells, two like electrodes at different concentrations are dipping in the same solution. Two hydrogen electrodes at unequal gas pressure immersed in the same solution of hydrogen ions constitute an electrode-concentration cell. This may be represented as follows:

$$Pt \mid H_2(\text{Pressure } P_1) : H^+ \mid\mid H^+ : H_2(\text{Pressure } P_2) \mid Pt$$

At left electrode (anode): $H_2(P_1) \longrightarrow 2H^+ + 2e^-$

At right electrode (cathode): $2H^+ + 2e^- \longrightarrow H_2(P_2)$

Cell reaction: $H_2(P_1) \longrightarrow H_2(P_2)$

It is clear that in this process there is no overall chemical change and there is only transfer of H_2 gas from the electrode with pressure P_1 to the electrode with P_2.The E.M.F. depends only on two pressures and independent on the concentration of electrolyte solutions hydrogen ions, in which the electrodes are dipped.

2. Electrolyte concentration cells. In these cells, the two electrodes of the same metal are dipped in solutions of metal ions of different concentrations, due to which a difference in the potential at the electrode and the electrolyte solution in which it is dipped. The source of electrical energy in the cell is the tendency of the electrolyte to diffuse from a solution of higher concentration to that of lower concentration.

A concentration cell with two zinc electrodes dipped in two solutions of $ZnSO_4$ with different concentrations joined through a salt bridge can be represented as follows:

$$Zn \mid Zn^{+2}(C_1) \mid\mid Zn^{+2}(C_2) \mid Zn \text{ (} C_2 \text{ is greater than } C_1\text{)}$$

The concentration ($C_2 > C_1$) the electrode reaction on the left is oxidation.

At left electrode (anode): $Zn \longrightarrow Zn^{+2}(C_1) + 2e^-$

At right electrode (cathode): $Zn^{+2}C_2 + 2e^- \longrightarrow Zn$

Cell reaction: $Zn^{+2}(C_2) \longrightarrow Zn^{+2}(C_1)$

The E.M.F. of such cell can be calculated by the following expression based on Nernst equation.

$$E = \frac{2.303\,RT}{2F}\log\frac{C_2}{C_1}$$

or

$$E = \frac{0.0592\ V}{2}\log\frac{C_2}{C_1}$$

Applications of concentration cells:

1. To determine the solubility of a sparingly soluble salt.
2. To calculate the valency of an ion.
3. To determine the transition point.
4. To calculate the extent of corrosion in metals.

2.14 ION-SELECTIVE ELECTRODE

Ion-selective electrode possesses the ability to respond only to certain specific ions, thereby developing a potential with respect to that species only in a mixture and ignoring the other ions totally. In other words, the potential developed by an ion-selective electrode depends only on the concentration of species/ions of interest. For example, glass membrane is only H^+ ion selective. The material employed for some ion-selective membranes are mentioned below.

Solid state membranes:

1. For fluoride ions-Lanthanum trifluoride crystal doped with Europium difluoride.
2. For chloride ions-pressed pellet of $Ag_2S + AgCl$. The electrode has a Teflon body and the pellet is held in position by using epoxy resin.

Liquid state membranes:

1. For certain alkali and alkaline-earth cations-neutral mono cyclic crown ethers and phosphate di ester.
2. For anions-Iron phenanthrolic complexes liquid state membranes usually obtained by absorbing the active molecules on an inert porous support such as porous polymer.

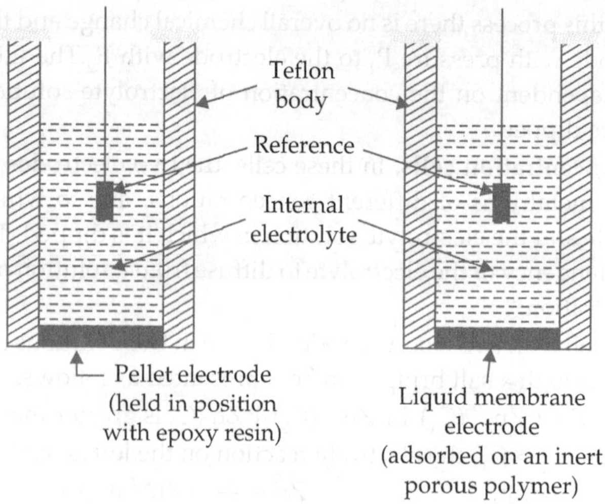

FIGURE 2.13 **Some ion selective electrodes.**

Determination of Concentration of Metal Ion. A schematic set up for determining the concentration of a metal ion of interest in a solution by using ion-selective electrode is

Reference electrode	Solution to be analysed $[M^{n+}] = C_1 = ?$	Ion-selective membrane	Internal standard solution $[M^{n+}] = C_2$	Identical reference electrode

This arrangement is similar to concentration cell and its potential is given by

$$E_{cell} = \frac{RT}{nF}\log\frac{C_2}{C_1} - \Delta E_{ref}$$

$$= \frac{RT}{nF}\log\frac{C_2}{C_1} \qquad [\Delta E_{ref} = 0 \text{ for identical reference electrode}]$$

$$= \frac{RT}{nF}\log C_2 - \frac{RT}{nF}\log C_1$$

$$= \text{Constant} - \frac{RT}{nF}\log C \quad [C = C_1]$$

$$\therefore \quad E_{cell} = \text{Constant} - \frac{0.0592\ V}{n}\log C \quad (\text{at } 25°C)$$

This equation is valid if the membrane employed is selective only to the ions to be analysed. However, the concentration of the other ions present in the system or mixture also have some effect on the value of membrane potential. Consequently, the above equation is modified to

$$E_{cell} = \text{Constant} - \frac{0.0592\ V}{n}\log C + \Sigma K_i C_i$$

where C_i = Concentration of the other ions present in the system/mixture

K_i = Selectivity coefficient of the ions.

Evidently, for a membrane to work more selectively, K_i values should be as small as possible.

Limitations:

1. Usually, ion-selective membranes can be used in solution of pH value up to 10, since higher pH values affect these.
2. Although ion selective membranes are very thin, yet their resistances are extremely high (of the order million ohms), so it is necessary to use electronic potentiometers to measure the potential difference.

Applications: In the determination of

1. Concentration of cations *e.g.*, H^+, Li^+, Na^+, K^+, NH_4^+, Ag^+, Pb^{+2}, Cd^{+2}, hardness (Ca^{+2}, Mg^{+2}) and anions, *e.g.*, halide ions (Cl^-, F^-, Br^-, I^-), NO_3^-, CN^-, S^{-2} etc.
2. pH of a solution by using H^+ ion-selective glass electrode.
3. Concentration of a gas by using gas sensing electrode. For example, glass electrode is employed for measuring the $CO_2(g)$ level in blood.

The glass electrode kept in contact with a very thin walled $CO_2(g)$ permeable silicone rubber membrane soaked in dilute $NaHCO_3$ solution. Then, the electrode is dipped in the blood sample under-rest. Since the membrane is permeable to $CO_2(g)$. Eventually, the pH value is changed, which is sensed by the glass electrode. In other words, observed potential of such a glass electrode can be employed to measure the concentration of $CO_2(g)$ in the blood sample.

Example for an Ion-Selective Electrode:

Glass electrode. When two solutions of different pH values are separated by a thin glass membrane, there develops a difference of potential between the two surfaces of the membrane. The potential difference developed is proportional to the difference in pH value. The glass membrane functions as an ion-exchange resin, and an equilibrium is set up between the Na^+ ions of glass and H^+ ions in solution. For a particular type of glass, the potential difference varies with the H^+ ion concentration, and is given by the expression

$$E_G = E_G^0 + 0.0592 \text{ V pH}$$

Over a range of pH of the rest-solution from 0 to 10.

Construction. A glass electrode consists of thin walled glass bulb containing AgCl-coated Ag electrode or simply a Pt electrode in 0.1 M HCl. The glass electrode may be shown schematically as,

$$-\text{Ag} \mid \text{AgCl}(s), \text{HCl (0.1 M)} \mid\mid \text{glass or Pt, (0.1 M) HCl} \mid \text{glass}^+$$

HCl in the bulb furnishes a constant H^+ ion concentration. Thus, it is an Ag-AgCl electrode, reversible with respect to chloride ions.

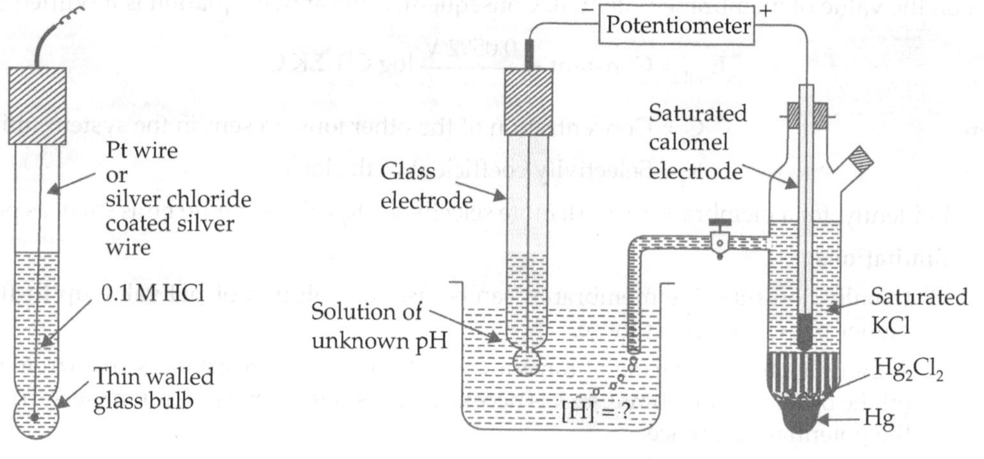

(*a*) Glass electrode (*b*) Determination of pH by glass electrode.

FIGURE 2.14

Glass electrode is used as the "Internal reference electrode" for determining the pH of solutions, especially coloured solutions containing oxidizing or reducing agents. Usually, Calomel electrode is used as the second electrode. In order to determine the pH of a solution, the glass electrode is placed in the solution under-rest and this half-cell is coupled with saturated calomel electrode. The e.m.f. of the cell is measured. Since the resistance is very high, so special electron tube voltammeters are used to measure the e.m.f. of the above cell. The e.m.f. of the complete cell is given by

$$E_{cell} = E_{Right} - E_{Left} = 0.2422 \text{ V} - [E_G^0 + 0.0592 \text{ V pH}]$$

$$\therefore \quad \text{pH} = \frac{0.2422 \text{ V} - E_{cell}}{0.0592 \text{ V}} = E_G^0$$

The E_G^0 value of a glass electrode can be determined by using a solution of known pH.

Advantages of glass electrode:

1. It is simple and can easily be used.
2. Equilibrium is rapidly achieved.
3. The results are accurate.
4. It is not easily poisoned.

Limitations:

1. The glass electrode can be used in solutions with pH range of 0 to 10. Electrodes composed of special glasses can be used for measurements up to a pH of 12. However, above 12 pH, cations of solution affect the glass interface and render the electrode useless.
2. Although glass membrane of electrode is very thin, yet its resistance is extremely high (of the order 10 to 100 million ohms), which cannot be measured by ordinary potentiometers. It is, therefore, necessary to use special electronic potentiometers.

2.15 TYPES OF ION-SELECTIVE ELECTRODE

Glass Membrane Electrodes. Glass membranes are made from an ion-exchange type of glass (silicate or chalcogenide). This type of ISE has good selectivity, but only for several single-charged cations; mainly H^+, Na^+, and Ag^+. Chalcogenide glass also has selectivity for double-charged metal ions, such as Pb^{2+}, and Cd^{2+}. The glass membrane has excellent chemical durability and can work in very aggressive media.

Construction. These electrodes are prepared from glass capillary tubing approximately 2 millimetres in diameter, a large batch at a time. Polyvinyl chloride is dissolved in a solvent and plasticizers (typically phthalates) added, in the standard fashion used when making something out of vinyl. In order to provide the ionic specificity, a specific ion channel or carrier is added to the solution; this allows the ion to pass through the vinyl, which prevents the passage of other ions and water.

One end of a piece of capillary tubing about an inch or two long is dipped into this solution and removed to let the vinyl solidify into a plug at that end of the tube. Using a syringe and needle, the tube is filled with salt solution from the other end, and may be stored in a bath of the salt solution for an indeterminate period. For convenience in use, the open end of the tubing is fitted through a tight o-ring into somewhat larger diameter tubing containing the same salt solution, with a silver or platinum electrode wire inserted. New electrode tips can thus be changed very quickly by simply removing the older electrode and replacing it with a new one.

Polymer Membrane Electrodes or Solid or Liquid Matrix Electrodes. Consist of various ion exchange materials in an inert matrix such as PVC, polyethylene or silicone rubber. The membrane is sealed to the end of a PVC tube. The potential developed at the membrane surface is related to the concentration of the species of interest. Electrodes of this type include Potassium, Calcium and Nitrate.

Solid State Electrodes. Utilize relatively insoluble inorganic salts in a membrane. Solid state electrodes exist in homogeneous or heterogeneous forms. In both types potentials are developed at the membrane surface due to the ion exchange process. Examples of this type of electrode include silver/sulfide, chloride and fluoride.

Gas Sensing Electrodes. Gas sensing electrodes are available for the measurement of ammonia, carbon dioxide, and nitrogen oxide. This type of electrode has a gas permeable membrane and an internal buffer solution. The pH of the buffer solution changes as the gas reacts with it. The change is detected by a combination pH sensor within the housing. This type of electrode does not require a an external reference electrode.

Glass Membrane Electrodes. Glass membrane electrodes are formed by the doping of the silicon dioxide glass matrix with various chemicals. The most common of the electrode of this type is the pH electrode. Glass membrane electrodes are also used for sodium ions.

Reference Electrodes. Ion Selective Electrodes are available as half-cells (mono) and combination electrodes where the reference is built into the probe. Available in glass or epoxy. Mono electrodes require the use of an additional reference electrode.

2.16 BATTERIES OR CELLS

"A device that stores chemical energy for later release as electricity is known as ***Battery***."

Or

An electrochemical cell or often, several electrochemical cells connected in series that can be used as source of direct electric current at a constant voltage.

Batteries or cells can be divided into three types:

1. Primary cells.
2. Secondary cells.
3. Fuel cells.

2.16.1 Primary Cells

Primary cells in which the cell reaction is not reversible. Thus, when the reactants have most of converted to products, no more electricity is produced and the battery or cell becomes dead.

Example. Dry cell or Leclanche cell.

(*i*) **Dry cell or Leclanche cell.** A cell without fluid component is perhaps the most familiar of all batteries. The anode of the cell is zinc (or container) containing an electrolyte consisting of NH_4Cl, $ZnCl_2$ and MnO_2 to which starch is added to make it thick paste-like so that it is less likely to leak. A carbon (graphite) rod serves as the cathode, which is immersed in the electrolyte in the centre as the cell. The anode (oxidation) half-cell reaction is

Oxidation: $Zn(s) \longrightarrow Zn^{+2}(aq) + 2e^-$ (at anode)

The cathode reaction is quite complex. Essentially, it involves the reaction of MnO_2 to a series of compounds having Mn in +3 oxidation state. For example, Mn_2O_3.

Reduction: $2MnO_2(s) + H_2O + 2e^- \longrightarrow Mn_2O_3(s) + 2OH^-(aq)$ (at cathode)

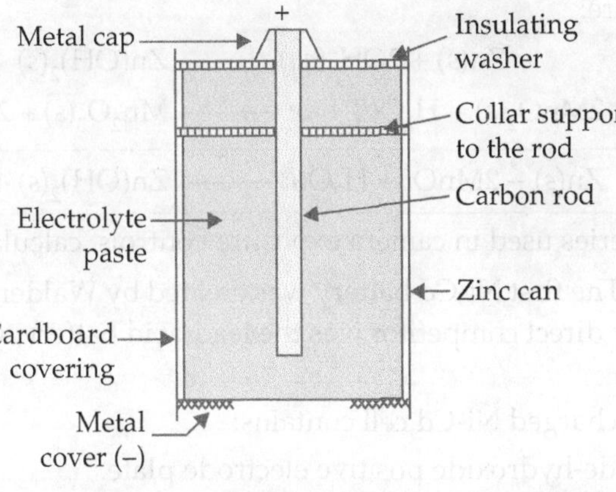

FIGURE 2.15 **Schematic diagram of a dry cell.**

However an acid-base reaction between OH^- and NH_4^+ (derived from NH_4Cl) evolves $NH_3(g)$,which disrupts the current flow.

$$2NH_4Cl(aq) + 2OH^-(aq) \longrightarrow 2NH_3(g) + 2H_2O + 2Cl^-$$

This is prevented by a reaction of $NH_3(g)$ with Zn^{+2}(from $ZnCl_2$) to form the complex ion $[Zn(NH_3)_2]Cl_2$. Hence, the reaction at two electrodes are,

$$Zn^{+2} + 2NH_3 + 2Cl^- \longrightarrow [Zn(NH_3)_2]Cl_2$$

Anode: $Zn(s) \longrightarrow Zn^{+2}(aq) + 2e^-$

Cathode: $2MnO_2(s) + 2NH_4Cl + Zn^{+2} + 2e^- \longrightarrow Mn_2O_3 + [Zn(NH_3)_2]\ Cl_2 + H_2O$

Overall reaction: $Zn(s) + 2MnO_2 + 2NH_4Cl \longrightarrow Mn_2O_3 + [Zn(NH_3)_2]\ Cl_2 + H_2O$

The dry cell is a primary cell, since various reactions involved cannot be reversed by passing electricity back through the cell. Dry cell is cheap to make, and gives a voltage of about 1.5 V. But it has two significant disadvantages.

- When current is drawn rapidly from it, products build up on the electrodes there by causing drop in voltage.
- Since the electrolytic medium is acidic, so zinc metal dissolves slowly, thereby the cell run down slowly, even if it is not in use.

Uses:

1. Dry cell is used in flash-lights.
2. Used in transistor radios.
3. Used in calculators etc.

(*ii*) **Alkaline battery.** The alkaline dry battery was invented by Canadian engineer Lewis Urry in the 1950s while working for the Everyday Battery company. On October 9, 1957, Lewis Urry, Karl Kordesch, and P.A. Marsal filed US patent (2,960,558) for the alkaline battery. It was granted in 1960 and was assigned to Union Carbide Corporation.

Chemistry. In an alkaline battery, the anode (negative terminal) is made of zinc powder, which gives more surface area for increased current, and the cathode (positive terminal) is composed of manganese dioxide. Unlike zinc-carbon (Leclanché) batteries, the electrolyte is potassium hydroxide rather than ammonium chloride or zinc chloride.

The half-reactions are:

Anode: $Zn(s) + 2OH^-(aq) \longrightarrow Zn(OH)_2(s) + 2e^-$

Cathode: $2MnO_2(s) + H_2O(l) + 2e^- \longrightarrow Mn_2O_3(s) + 2OH^-(aq)$

Net reaction: $Zn(s) + 2MnO_2 + H_2O(l) \longrightarrow Zn(OH)_2(s) + Mn_2O_3(s)$

Uses. Alkaline batteries used in camera exposure controls, calculators, watches, etc.

(*iii*) **Ni-Cd battery.** The first Ni-Cd battery was created by Waldemar Jungner of Sweden in 1899. At that time, the only direct competitor was the lead-acid battery, which was less physically and chemically robust.

Chemistry. A fully charged Ni-Cd cell contains:

- A nickel (III) oxide-hydroxide positive electrode plate.
- A cadmium negative electrode plate.
- A separator.
- And an alkaline electrolyte (potassium hydroxide).

The chemical reactions in a Ni-Cd battery during discharge are:

Anode: $Cds(s) + 2OH^-(aq) \longrightarrow Cd(OH)_2(s) + 2e^-$

Cathode: $2NiO(OH)(s) + 2H_2O(l) + 2e^- \longrightarrow 2Ni(OH)_2(s) + 2OH^-(aq)$

Net reaction: $2NiO(OH)(s) + Cd(s) + 2H_2O(l) \longrightarrow Cd(OH)_2(s) + 2Ni(OH)_2(s)$

The reaction can be readily reversed, because the reaction products, $Ni(OH)_2$ (*s*) and $Cd(OH)_2(s)$, adhere to the electrode surfaces.

Ni–Cd Battery, fairly recently developed, is a portable, rechargeable cell and its cell voltage is fairly constant (about 1.4 V). Like a dry cell, it can be packed in a sealed container. It can be left for long periods of time without any appreciable deterioration, since no gases are produced during discharging (or charging).

Uses. It is used in electronic calculators, electronic flash units, cordless electronic shavers, transistors and other battery powered small tools.

(*iv*) **Mercury battery.** A mercury battery (also called mercuric oxide battery or mercury cell) is a non-rechargeable electrochemical battery, a primary cell. This battery first appeared in the early 1940s. Due to the content of mercury, and the resulting environmental concerns, the sale of mercury batteries is banned in many countries.

The mercury battery has many applications in medicine and electronics. The battery consists of a steel—made container in the shape of a cylinder acting as the cathode, where an amalgamated anode of mercury and zinc is surrounded by a stronger alkaline electrolyte and a paste of zinc oxide and mercury (II) oxide.

Mercury battery half reactions are shown below:

Anode: $Zn(Hg) + 2OH^-(aq) \longrightarrow ZnO(s) + H_2O(l) + 2e^-$

Cathode: $HgO(s) + H_2O(l) + 2e^- \longrightarrow Hg(l) + 2OH^-(aq)$

Overall reaction: $Zn(Hg) + HgO(s) \longrightarrow ZnO(s) + Hg(l)$

Uses. It gives excellent performance in heart pacemakers, hearing acids, light meters and digital watches.

(*v*) **Lithium cells.** A lithium-ion battery (sometimes Li-ion battery or LIB) is a family of rechargeable battery types in which lithium ions move from the negative electrode to the positive electrode during discharge, and back when charging. Chemistry, performance, cost, and safety characteristics vary across LIB types. Unlike lithium primary batteries (which are disposable), lithium-ion electrochemical cells use an intercalated lithium compound as the electrode material instead of metallic lithium.

Both the anode and cathode are materials into which, and from which, lithium can migrate. During *insertion* (or *intercalation*) lithium moves into the electrode. During the reverse process, extraction (or *deintercalation*), lithium moves back out. When a lithium-based cell is discharging, the lithium is extracted from the anode and inserted into the cathode. When the cell is charging, the reverse occurs.Usefulwork can only be extracted if electrons flow through a closed external circuit. The following equations are in units of moles, making it possible to use the coefficient X.

The positive electrode half-reaction (with charging being forwards) is

$$LiCoO_2 \rightleftharpoons Li_{1-x}CoO_2 + xLi^+ + xe^-$$

The negative electrode half-reaction is

$$xLi^+ + xe^- + 6C \rightleftharpoons Li_xC_6$$

The overall reaction has its limits. Over discharge supersaturates lithium cobalt oxide, leading to the production of lithium oxide, possibly by the following irreversible reaction:

$$Li^+ + LiCoO_2 \longrightarrow Li_2O + CoO$$

Overcharge up to 5.2 volts leads to the synthesis of Cobalt (IV) oxide, as evidenced by X-ray diffraction

$$LiCoO_2 \longrightarrow Li^+ + CoO_2$$

In a lithium-ion battery the lithium ions are transported to and from the cathode or anode, with the transition metal, cobalt (Co), in Li_xCoO_2 being oxidized from Co^{3+} to Co^{4+} during charging, and reduced from Co^{4+} to Co^{3+} during discharge.

Uses. Lithium batteries can be used in place of ordinary alkaline cells in many devices, such as clocks and cameras. Lithium batteries find application in many long-life, critical devices, such as artificial pacemakers and other implantable electronic medical devices. These devices use specialized lithium-iodide batteries designed to last 15 or more years. But for other, less critical applications such as in toys, the lithium battery may actually outlast the device. In such cases, an expensive lithium battery may not be cost-effective.

(*vi*) **Zn-C battery.** A zinc-carbon dry cell or battery is packaged in a zinc can that serves as both a container and negative terminal. It was developed from the wet Leclanché cell. A zinc-carbon dry cell is considered as a primary cell because the cell is not intended to be recharged.

The positive terminal is a carbon rod surrounded by a mixture of manganese dioxide and carbon powder. The electrolyte used is a paste of zinc chloride and ammonium chloride dissolved in water. Zinc chloride cells are an improved version from the original ammonium chloride variety. Zinc-carbon batteries are the least expensive primary batteries and thus a popular choice by manufacturers when devices are sold with (low quality) batteries included. They are commonly labelled as "General Purpose" batteries.

Chemical reactions. In a zinc-carbon dry cell, the outer zinc container is the negative terminal. The zinc is oxidised according to the following half-equation.

At anode: $Zn(s) \longrightarrow Zn^{2+}(aq) + 2e^-$

At cathode: $2MnO_2(s) + 2e^- + 2NH_4Cl(aq) \longrightarrow Mn_2O_3(s) + 2NH_3(aq) + H_2O(aq) + 2Cl^-$

At net reaction: $Zn(s) + 2MnO_2(s) + 2NH_4Cl(aq) \longrightarrow Mn_2O_3(s) + Zn(NH_3)_2Cl_2(aq) + H_2O(l)$

The battery has an e.m.f. of about 1.5 V. The approximate nature of the e.m.f. is related to the complexity of the cathode reaction. The anode (zinc) reaction is comparatively simple with a known potential. Side reactions and depletion of the active chemicals increases the internal resistance of the battery, and this causes the e.m.f. to drop.

Although carbon is an important element of the battery's construction, it takes no part in the electrochemical reaction, instead only serving to collect current and reduce the resistance of the manganese dioxide mix. The cell could more properly be called a "zinc-manganese" cell.

Uses. They were the first commercial dry battery and made things such as flashlights and other portable devices possible, because the battery can function in any position. They can be used in remote controls, flashlights, clocks, or transistor radios, since the power drain is not too heavy (suited for low drain use). They are progressively replaced, in many usages, by alkaline cells, and rechargeable NiMH batteries (which of course did not exist at Leclanché's time).

2.16.2 Secondary Cells

Secondary cells in which the cell reaction can be reversed by passing direct electric current in opposite direction. Thus, a secondary cell may be used through a large number of cycles of discharging and charging.

Example. Lead–Acid storage cell.

Lead –Acid storage cell. A storage cell is one that can operate both as a voltaic cell and as an electrical cell. When operating as a voltaic cell, it supplies electrical energy and as a result eventually becomes "rundown". It must then be recharged. When being recharged, the cell operates as an electrolytic cell. Thus, storage cell has the great advantage of its ability to work both ways, to receive electrical energy and also to supply it.

The common example of a storage cell is the lead –acid storage cell. One of its electrodes is made of lead. The other electrode is made of lead dioxide (PbO_2) or rather a paste of PbO_2 is pressed into a grid, made of lead. A number of lead plates (–ve plates) are connected in parallel and a number of lead dioxide plates (+ve plates) are also connected in parallel. The lead plates fit in–between the lead dioxide plates. Various plates are separated from the adjacent ones by insulators like strips of wood or rubber or glass fiber. The entire combination is then immersed in approximately 20 to 21% dil . H_2SO_4 corresponding to a density of 1.2 to 1.3.

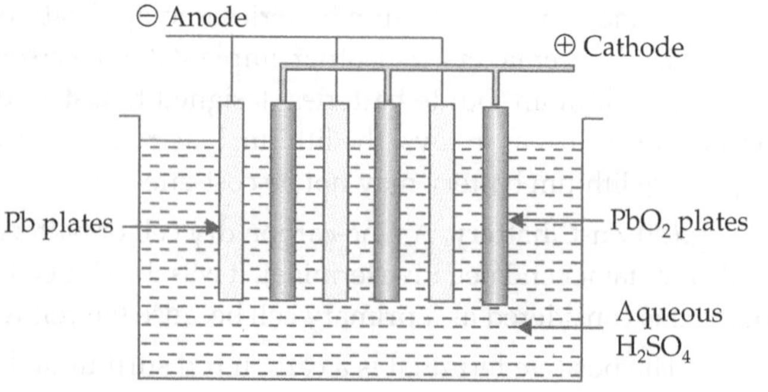

FIGURE 2.16 **Lead storage cell.**

Discharging. When a storage cell is operating as a voltaic cell, it is said to be discharging. The lead electrode loses electrons, which flow through the wire. In this reaction, oxidation of lead takes place at the anode.

$$Pb \longrightarrow Pb^{+2} + 2e^- \quad \text{(at lead anode)}$$

The Pb^{+2} ions then combine with sulphate (SO_4^{-2}) ions.

$$Pb^{+2} + SO_4^{-2} \longrightarrow PbSO_4\downarrow$$

The electrons released from the anode (lead plate) flows to the dioxide electrode. Here PbO_2 gains electrons to form Pb^{+2} ions. In other words, lead undergoes reduction at the cathode from oxidation state +4 to +2. The Pb^{+2} ions then combine with SO_4^{-2} ions.

$$PbO_2 + 4H^+ + 2e^- \longrightarrow Pb^{+2} + 2H_2O \quad \text{(at } PbO_2 \text{ cathode)}$$

$$Pb^{+2} + SO_4^{-2} \longrightarrow PbSO_4$$

So the net reaction during use (or discharging) is,

$$Pb + PbO_2 + 4H^+ + 2SO_4^{-2} \longrightarrow 2PbSO_4 + 2H_2O + \text{Energy}$$

It may be noted that lead sulphate is precipitated at both the electrodes. The voltage of each cell is about 2.0 volts at a concentration of 21.4% H_2SO_4 at 25°C. Lead cell, commonly used in automobiles, is a combination of six such cells in series to form a battery with an e.m.f. of 12 volts.

Charging. When both anode and cathode become covered with $PbSO_4$, the cell ceases to function as a voltaic cell. To recharge a lead storage cell, the reactions taking place during discharging are reversed by passing an external e.m.f. greater than 2 volts from a generator. The +ve pole of the generator is attached to the –ve pole the cell and the following reactions take place.

Reaction at the –ve terminal (cathode)

$$PbSO_4 + 2e^- \longrightarrow Pb + SO_4^{-2}$$

Reaction at the +ve terminal (anode)

$$PbSO_4 + 2H_2O - 2e^- \longrightarrow PbO_2 + 4H^+ + SO_4^{-2}$$

Hence, the net reaction during charging is

$$2PbSO_4 + 2H_2O + \text{Energy} \longrightarrow Pb + PbO_2 + 4H^+ + 2SO_4^{-2}$$

During the process the charging, the electrodes of cell are restored to their original conditions (*i.e.*, Pb and PbO_2 respectively). It may also be noted that during discharging operation, the concentration of acid (H_2SO_4) decreases, while the concentration of acid increases during the charging operation.

Uses. Lead acid storage cells are used for many purposes such as to supply current for electrical vehicles, gas engine ignition, in telephone exchanges, railways, mines, laboratories, hospitals, broadcasting stations, automobiles, power stations and distribution works.

Note:

1. The lead storage cell is both a voltaic cell and an electrolytic cell. When electricity is being drawn from the cell to start the car, it acts as a voltage cell. The alternator forces electrical energy (generated by it due its motion) into the cell, thereby causing electrolysis reactions to take place that restore the original reactants of the cell.
2. The lead storage battery (of 12 V) commonly used in automobiles consists of six identical cell joined together in series.

2.16.3 Fuel Cells

"A device for converting the energy of a fuel directly into electrical energy is known as a ***fuel cell***".

We know that voltaic cells involve oxidation–reduction reaction. A workable cell forces the electrons that leave the substance being oxidized through an external conductor before they are received by the oxidizing agent. We also know that all ordinary combustion reactions are redox reactions. However, when a fuel is burnt, the electron exchange takes place only when the atoms of the oxidizing agent (oxygen or air) come in direct contact with the atoms of the substance being oxidized (the fuel). The energy is released as heat. In a fuel cell, electric energy is obtained without combustion from oxygen and a gas that can be oxidized. Hence, a fuel cell converts chemical energy of the fuel cells directly to electricity. The process in fuel cell is,

$$\text{Fuel + Oxygen} \longrightarrow \text{Oxidation products + Electricity}$$

Example. Hydrogen-Oxygen fuel cell.

Hydrogen-oxygen fuel cell. One of the simplest and most successful fuel cell is Hydrogen –oxygen fuel cell. It consists essentially of an electrolytic solution such as 25% KOH solution, and two inert porous electrodes. Hydrogen and oxygen gas are bubbled through the anode and cathode compartment respectively, where the following reactions take place.

Anode: $2H_2(g) + 4OH^-(aq) \longrightarrow 4H_2O(l) + 4e^-$

Cathode: $O_2(g) + 2H_2O(l) + 4e^- \longrightarrow 4OH^-(aq)$

Net reaction: $2H_2(g) + O_2(g) \longrightarrow 2H_2O(l)$

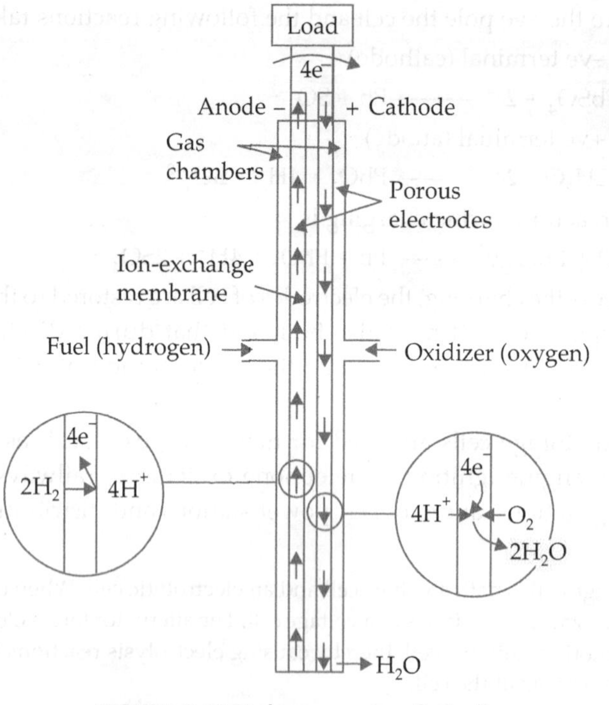

FIGURE 2.17 **Hydrogen-oxygen fuel cell.**

The standard e.m.f. of the cell

$$E^0 = E_{oxd} + E_{red} = 0.83\text{ V} + 0.40\text{ V} = 1.23\text{ V}$$

In actual practice, the e.m.f. of the cell is 0.8 to 1.0 V. It may be note that the product discharged by the cell is water. Usually, a large number of these cells are stacked together in series to make a battery called ***fuel battery***.

It may be pointed have that the electrodes must meet stringent requirements, they must

- Be good conductors,
- Be good electron-sources or sinks, and
- Not be consumed or deteriorated by the electrolyte heat or electrode reactions.

SOLVED PROBLEMS

Problem 1. *The resistance of 0.5N solution of an electrolyte in a cell was found to be 50 Ω. Calculate the equivalent conductance of the solution if the electrode in the cell are 2.2 cm apart with an area of 3.8 cm².*

Solution. Length of the electrodes = 2.2 cm.

Area of cross-section of electrodes = 3.8 cm²

$$\text{Cell constant (X)} = \frac{l}{s} = \frac{2.2}{3.8} = 0.5789 \text{ cm}^{-1}$$

Observed resistance R = 50 Ω

$$\text{Observed conductance C} = \frac{1}{R} = \frac{1}{50} = 0.02\ \Omega^{-1}$$

$$\text{Specific conductance } (k) = X \times \text{Observed conductance}$$
$$= 0.5789 \times 0.02$$
$$= 0.01158\ \Omega^{-1}\text{ cm}^{-1}$$

$$\text{Equivalent conductance } \lambda_{eq} = k \times v$$

$$\lambda_{eq} = k \times \frac{1000}{N} = \frac{0.01158 \times 1000}{0.5}$$
$$= 3.16\ \Omega^{-1}\text{ cm}^2\text{ eq}^{-1}.$$

Problem 2. *The molar conductivity at infinite dilution of $AgNO_3$, NaCl and $NaNO_3$ is 116.51103 and 105.2 Ω^{-1} cm² mol⁻¹ respectively. The specific conductivity of AgCl is $1.24 \times 10^{-6}\ \Omega^{-1}\ cm^{-1}$. Find out the solubility of AgCl in g/litre.*

Solution.
$$\lambda_{eq}^{\infty} = k \times v$$

$$v = \frac{\lambda_{eq}^{\infty}}{k}$$

Solubility
$$S = \left(\frac{1000}{v}\right) \text{g eq/L} = \frac{1000\text{ E}}{v}\text{ g/L} = \frac{1000 \times E \times k}{\lambda_{eq}^{\infty}}\text{ g/L}$$

$$S = \frac{1000 \times 1.24 \times 10^{-6} \times 143.5}{121.6}\text{ g/L} \qquad [\because \text{ eq.wt of AgCl} = 143.5]$$

$$S = 1.4633 \times 10^{-3}\text{ g/L.}$$

Problem 3. *At 291 K the ionic velocities of Ag^+ is 0.00057 cm/s and that of NO_3^- is 0.00063 cm/s. What is the value of λ_∞ for $AgNO_3$? If the specific conductivity of 0.1 N $AgNO_3$ solution is 0.00947 $\Omega^{-1}\ cm^{-1}$, find the degree of ionization at this dilution.*

Solution.

$$\text{Ionic mobility} = \frac{\text{Ionic conductance}}{96500}$$

$$\lambda_{Ag^+} = 0.00057 \times 96500$$

$$\lambda_{NO_3^-} = 0.00063 \times 96500$$

$$\lambda_{AgNO_3} = \lambda_{Ag^+} + \lambda_{NO_3^-}$$

$$= 0.00057 \times 96500 + 0.00063 \times 96500$$

$$= 96500\ (0.00057 + 0.00063)$$

$$= 96500\ (0.0012)$$

$$= 115.8$$

$$\lambda_c = \frac{1000\ K}{C}$$

$$\lambda_c = \frac{1000 \times 0.00947}{0.1}$$

$$\lambda_c = 94.7$$

$$\text{Degree of ionization } \alpha = \frac{\lambda_c}{\lambda_\infty}$$

$$= \frac{94.7}{115.8} = 0.817$$

■ **Problem 4.** *Calculate the electrode potential of single electrode Zn^{+2}/Zn (0.1M). ($E^0 = -0.762$ V)*

Solution. $Zn^{+2} + 2e^- \rightleftharpoons Zn$

$$n = 2 \qquad \text{(Number of electrons)}$$

$$E = E^0 + \frac{0.059}{n} \log C \qquad \text{(for cation electrodes)}$$

$$E = -0.762 + \frac{0.059}{2} \log (0.1)$$

$$E = -0.762 + 0.0295 \log (10^{-1})$$

$$E = -0.762 - 0.0295$$

$$E = -0.792 \text{ V}$$

■ **Problem 5.** *Calculate the electrode potential of single electrode $Ag^+(0.01M)/Ag$. ($E^0 = 0.799$ V)*

Solution. $Ag^+ + 1e^- \rightleftharpoons Ag$

$$E = E^0 + \frac{0.059}{n} \log C \quad \text{(for cation electrodes)}$$

$$E = 0.799 + \frac{0.059}{1} \log (0.01)$$

$$E = 0.799 + 0.059 \log (10^{-2})$$

$$E = 0.799 + 0.059\ (-2)$$

$$E = 0.799 - 0.118$$

$$E = 0.681 \text{ V}$$

Problem 6. *Calculate the electrode potential of single electrode Pt, $Cl_2/Cl^-(0.01M)$. ($E^0 = +1.36$ V).*

Solution. $\frac{1}{2}Cl_2 + 1e^- \rightleftharpoons Cl^-$

$$E = E^0 - \frac{0.059}{n} \log C \quad \text{(for anion electrodes)}$$

$$E = 1.36 - \frac{0.059}{1} \log(0.01)$$

$$E = 1.36 - 0.059 \log(10^{-2})$$

$$E = 1.36 - 0.059(-2)$$

$$E = 1.36 + 0.118 = 1.478 \text{ V}$$

Problem 7. *Calculate the E.M.F. of $Zn/Zn^{+2}(0.1M)//Cu^{+2}(0.1M)/Cu$ [E^0 of $Zn^{+2}/Zn = -0.762$ V; E^0 of $Cu^{+2}/Cu = +0.337$ V].*

Solution.

$$E^0_{Zn^{+2}/Zn} = E_{left} = E^0 + \frac{0.059}{n} \log C$$

$$E_{left} = -0.762 + \frac{0.059}{2} \log(0.1)$$

$$E_{left} = -0.762 + 0.0295 \log(10^{-1})$$

$$E_{left} = -0.762 - 0.0295$$

$$E_{left} = -0.792 \text{ V}$$

$$E^0_{Cu^{+2}/Cu} = E_{right} = E^0 + \frac{0.059}{n} \log C$$

$$E_{right} = 0.337 + \frac{0.059}{2} \log(0.1)$$

$$E_{right} = 0.337 + 0.0295 \log(10^{-1})$$

$$E_{right} = 0.337 + 0.0295(-1)$$

$$E_{right} = 0.307 \text{ V}$$

$$E_{cell} = [E_{right} - E_{left}] = [0.307 - (-0.792)]$$

$$E_{cell} = 1.099 \text{ V}$$

Total E.M.F. of the cell is $E_{cell} = 1.099$ V

Problem 8. *A Nickel electrode ($E^0 = -0.13$ volt) is combined with an Aluminium electrode (standard oxidation potential is 1.66 volts) to form a cell. What would be the potential of the cell?* (SET-3, Feb 2012)

Solution. The above electrochemical cell is

$Al/Al^{+3}//Ni^{+2}/Ni$

E^0 of $Ni^{+2}/Ni = -0.13$ V; E^0 of $Al^{+3}/Al = +1.66$ V

$$E_{cell} = [E_{right} - E_{left}]$$

$$E_{cell} = [-0.13 - 1.66]$$

$$E_{cell} = -1.79 \text{ V}$$

Problem 9. *Calculate the standard electrode potential Cu^{++}/Cu, if the electrode potential at 25°C is 0.296 volt when $[Cu^{++}]$ is 0.015 M.* (SET-1, Feb 2011)

Solution.

$$E = E^0 + \frac{0.059}{n} \log C$$

$$0.296 = E^0 + \frac{0.059}{2} \log 0.015$$

$$0.296 = E^0 + \frac{0.059}{2} (-1.82)$$

$$0.296 = E^0 - 0.05$$

$$E^0 = 0.296 + 0.05$$

$$E^0 = 0.346 \text{ V.}$$

EXERCISE

1. Define the following:
   (*i*) Ohm's law
   (*ii*) Specific resistance
   (*iii*) Conductance
   (*iv*) Specific conductivity
   (*v*) Equivalent conductivity
   (*vi*) Molar conductivity.
2. Explain about the concept of ionic mobility.
3. State Kohlrausch's law. Explain about the applications of Kohlrausch's law.
4. Write a note on Conductometric titrations.
5. Explain about Galvanic cells.
6. Define electrode potential. How do you measure an electrode potential of a cell?
7. Derive Nernst equation.
8. Write a short note on Electrochemical series.
9. Explain about potentiometric titrations.
10. Define concentration cells. Mention and explain about their types.
11. What is an Ion-selective electrode? Explain with an example.
12. What is battery or cell? Explain the following with an example:
    (*a*) Primary cells
    (*b*) Secondary cells
    (*c*) Fuel cells.

Chapter 3

CORROSION

3.1 INTRODUCTION

Corrosion. Any process of destruction (or deterioration) and consequent loss of a solid metallic material, through an unwanted chemical or electrochemical attack by its environment, starting at its surface, is called corrosion.

The most familiar example of corrosion is rusting or iron, when exposed to the atmospheric conditions. During this, a layer of reddish scale and powder of oxide (Fe_3O_4) is formed, and the iron becomes brittle. Another common example is formation of green film of basic carbonate [$CuCO_3 + Cu(OH)_2$] on the surface of copper, when exposed to moist-air containing carbon dioxide.

Causes and Effects of Corrosion

Cause of corrosion. Metals exist in the form of oxides, sulphides, sulphates and carbonates. The chemically combined states of metal known as ore has low energy and is thus thermodynamically stable state for metal. The metals are extracted from these ores after supplying a large amount of energy. Metals in the uncombined condition have a higher energy and are in an unstable state. It is their natural tendency to go back to the low energy state *i.e.*, combined state by recombining with the elements present in the environment. This is the main reason for corrosion.

$$\underset{\text{(Thermodynamically Stable state)}}{\text{Ore of metal}} \xrightarrow[\text{Operation + Energy}]{\text{Metallurgical}} \underset{\text{(Thermodynamically unstable state)}}{\text{Pure metal}} \xrightarrow{\text{Environment}} \underset{\text{(Comparatively more stable than pure metal)}}{\text{Corroded metal}}$$

Although corroded metal is thermodynamically more stable than pure metal but due to corrosion useful properties of a metal such as malleability, ductility and electrical conductivity are lost.

Effects of corrosion: The effects of corrosion are:

1. Loss of useful properties of metal and thus loss of efficiency.
2. Replacement of corroded equipment.
3. High maintenance costs such as repainting.
4. Contamination of the product.
5. Due to failure, the plant shuts down.
6. Pollution due to a corrosion product affects the health.

Corrosion can be prevented by:

1. Proper design and use of suitable material.
2. Cathodic and anodic protection.
3. Treatment of metals.
4. External influences, and
5. Environmental modifications.

3.2 DRY OR CHEMICAL CORROSION

This type of corrosion occurs mainly through the direct chemical action of environment/atmospheric gases such as oxygen, halogen, hydrogen sulphide, sulphur dioxide, nitrogen or anhydrous inorganic liquid with metal surfaces in immediate proximity. There are three main types of chemical corrosions.

1. Oxidation corrosion 2. Corrosion by other gases 3. Liquid metal corrosion.

3.2.1 Oxidation Corrosion

Oxidation corrosion is brought about by the direct action of oxygen at low or high temperatures on metals, usually, in the absence of moisture. At ordinary temperatures, metals, in general, are very slightly attacked. However, alkali metals (Li, Na, K, Rb, etc.) and alkaline-earth (Be, Ca, Sr, etc.) are even rapidly oxidized at low temperatures. At high temperatures, almost all metals (except Ag, Au, and Pt) are oxidized.

The reaction in the oxidation corrosion are:

$$2M \longrightarrow \underset{\text{Metal ions}}{2M^{n+}} + 2ne^- \quad \text{(Loss of electrons)}$$

$$\frac{n}{2}O_2 + 2ne^- \longrightarrow \underset{\text{Oxide ions}}{nO^{2-}} \quad \text{(Gain of electrons)}$$

$$2M + \frac{n}{2}O_2 \longrightarrow \underbrace{\underset{\text{Metal ions}}{2M^{n+}} + \underset{\text{Oxide ions}}{nO^{2-}}}_{\text{Metal oxide}}$$

Mechanism. Oxidation occurs first at the surface of the metal and the resulting metal oxide scale forms a barrier that tends to restrict further oxidation. For oxidation to continue either the metal must diffuse outwards through the scale to the surface or the oxygen must diffuse inwards through the scale to the underlying metal. Both transfers occur, but the outward diffusion of metal is, generally, much more rapid than the inward diffusion of oxygen, since the metal ion is appreciably smaller than the oxygen iron and consequently, of much higher mobility.

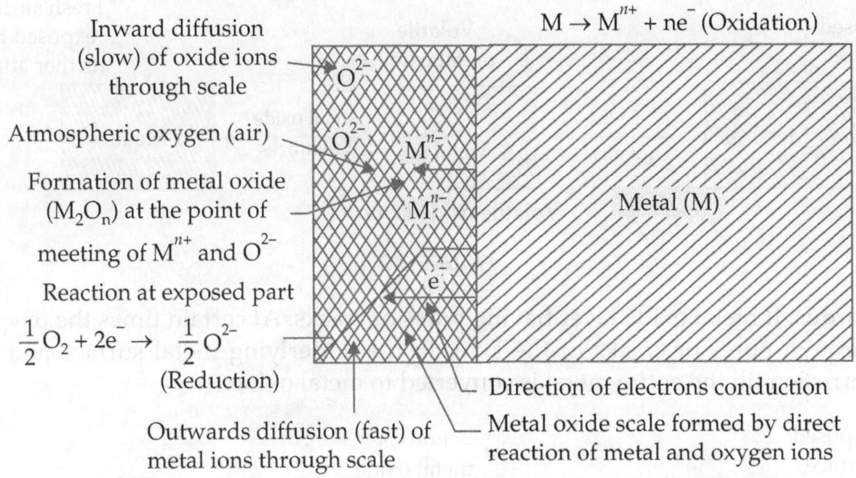

FIGURE 3.1

Nature of the oxide formed plays an important part in oxidation corrosion process.

$$\text{Metal} + \text{Oxygen} \longrightarrow \text{Metal oxide} \quad \text{(Corrosion product)}$$

When oxidation starts, a thin layer of oxide is formed on the metal surface and the nature of this film decides the further action. If the film is:

(*i*) **Stable:** The corrosion forms stable oxide film which is impervious or tightly adhering as seen in Cu or Al. This impervious coating will prevent further corrosion.

(*ii*) **Unstable:** Sometimes the oxide film decomposes back to metal and oxygen. In such conditions the corrosion is not possible. This is seen mostly in Ag, Al, and Pt. *i.e.*, the oxide layer formed decomposes back into the metal and oxygen.

$$\text{Metal oxide} = \text{Metal} + \text{Oxygen}$$

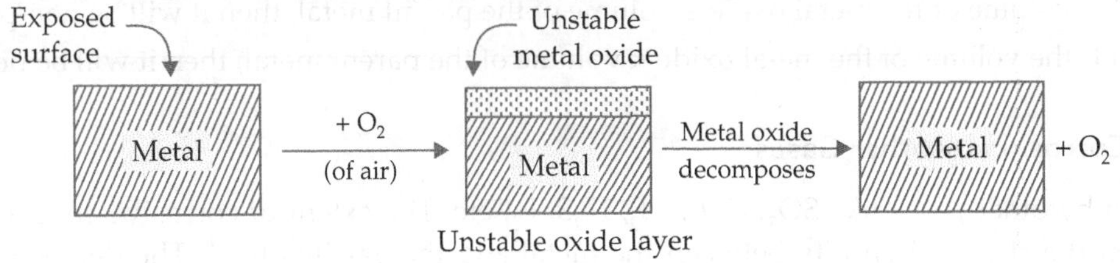

FIGURE 3.2

(*iii*) **Volatile:** In some metal surfaces, the oxide film gets volatized as soon as it is formed. This condition exposes the metal surface again and increases corrosion rapidly. Mostly seen in molybdenum.

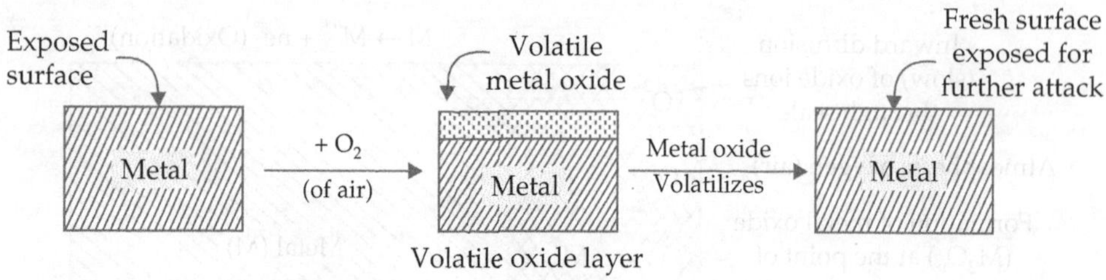

FIGURE 3.3

(*iv*) **Porous:** If metal oxide layer having pores or cracks. At certain times the oxygen passes through pores or cracks of the layer to the underlying metal surface and continues corrosion till the entire metal is converted to metal oxide.

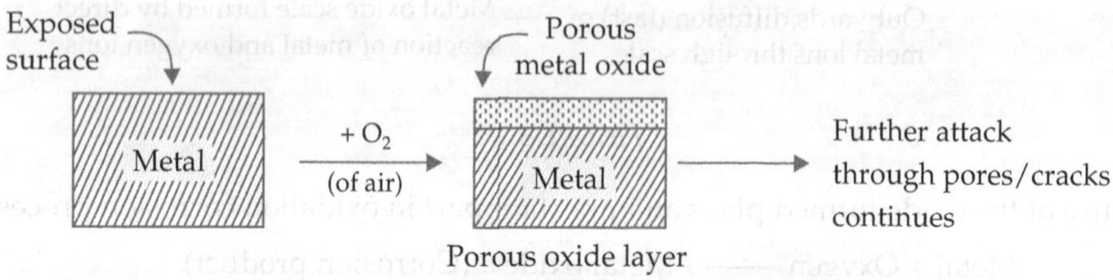

FIGURE 3.4

Pilling-Bed worth rule: Pilling-Bed worth rule describes the porous and non-porous nature of metallic oxide which is formed during corrosion.

According to this rule the specific volume ratio is calculated as follows:

$$\text{Specific volume ratio} = \frac{\text{Volume of metal oxide}}{\text{Volume of metal}}$$

(*a*) If the specific volume ratio is smaller, then oxidation corrosion will takes place because the oxide films will be sufficiently porous for diffusion of M^{n+} and O^{2-}.

(*b*) If volume of the metal oxide $\geq$ volume of the parent metal, then it will be non-porous.

(*c*) If the volume of the metal oxide $<$ volume of the parent metal, then it will be porous.

3.2.2 Corrosion by other Gases

Corrosion by other gases like SO_2, CO_2, Cl_2, H_2S, F_2, etc. The extent of corrosive effect depends mainly on the chemical affinity between the metal and the gas involved. The degree of attack depends on the formation of protective or non-protective films on the metal surface.

(*a*) If the film formed is protective or non-porous (*e.g.*, AgCl film, resulting from the attack of Cl_2 on Ag), the intensity or extent of attack decreases, because the film formed protects the metal from further attack.

(*b*) If the film formed is non-protective or porous, the surface of the whole metal is gradually destroyed. For example, dry Cl_2 gas attacks on tin (Sn) forming volatile $SnCl_4$ (which volatilizes immediately), thereby leaving fresh metal surface for further attack. Similarly, in petroleum industry, H_2S at high temperature attacks steel forming a FeS scale, which is porous and interferes with normal operations.

3.2.3 Liquid Metal Corrosion

Liquid metal corrosion is due to chemical action of flowing liquid metal at high temperatures on solid metal or alloy. Such corrosion occurs in devices used for nuclear power. The corrosion reaction involves either: (*i*) dissolution of a solid metal by a liquid metal or (*ii*) internal penetration of the liquid metal into the solid metal. Both these modes of corrosion cause weakening of the solid metal.

3.3 WET OR ELECTROCHEMICAL CORROSION

This type of corrosion occurs:

(*i*) when a conducting liquid is in contact with metal or

(*ii*) when two dissimilar metals or alloys are either immersed or dipped partially in a solution.

This corrosion occurs, due to the existence of separate 'anodic' and 'cathodic' areas/parts, between which current flows through the conducting solution. At anodic area, oxidation reaction (*i.e.*, liberation of free electrons) takes place, so anodic metal is destroyed by either dissolving or assuming combined state (such as oxide, etc.) Hence, corrosion always occurs at anodic areas.

At anode: $\underset{\text{(Metal)}}{M} \longrightarrow \underset{\text{(Metal ion)}}{M^{n+}} + ne^-$ (Oxidation)

$\underset{\text{(Metal ion)}}{M^{n+}} \longrightarrow$ Dissolves in solution

$\quad\longrightarrow$ Forms compound such as oxide

On the other hand, at cathodic areas, reduction reaction (*i.e.*, gain of electrons) takes place. Usually, cathode reactions do not affect the cathode, since most metals cannot be further reduced. So at cathodic part, dissolved constituents in the conducting medium accept the electrons to form some ions (like OH^-, O^{2-}). The metallic ions (at anodic part) and non-metallic ions (formed at cathodic part) diffuse towards each other through conducting medium and form a corrosion product somewhere between anode and cathode. The electrons set free at the anode flow through the metal and are finally consumed in the cathodic reaction. Thus, we may sum up that electrochemical corrosion involves:

(*i*) The formation of anodic and cathodic areas or parts in contact with each other,

(*ii*) Presence of a conducting medium,

(*iii*) Corrosion of anodic areas only, and

(*iv*) Formation of corrosion product somewhere between anodic and cathodic areas.

Mechanism. On the other hand, the cathodic reaction consumes electrons with either by: (*a*) evolution of hydrogen or (*b*) absorption of oxygen, depending on the nature of the corrosive environment.

(*a*) **Evolution of hydrogen-type:** Corrosion occurs, usually, in acidic environments. Considering metal like Fe, the anodic reaction is dissolution of iron as ferrous ions with the liberation of electrons.

$$Fe \longrightarrow Fe^{2+} + 2e^- \quad \text{(Oxidation)}$$

These electrons flow through the metal, from anode to cathode, where H^+ ions (of acidic solution) are eliminated as hydrogen gas.

$$2H^+ + 2e^- \longrightarrow H_2\uparrow \quad \text{(Reduction)}$$

The overall reaction is: $Fe + 2H^+ \longrightarrow Fe^{2+} + H_2\uparrow$

Thus, this type of corrosion causes "displacement of hydrogen ions from the acidic solution by metal ions". Consequently all metals above hydrogen in the electrochemical series have a tendency to get dissolved in acidic solution with simultaneous evolution of hydrogen.

(*b*) **Absorption of oxygen:** Rusting of iron in neutral aqueous solution of electrolytes (like NaCl solution) in presence of atmospheric oxygen is a common example of this type of corrosion. The surface of iron is, usually coated with a thin film of iron oxide. However, if this iron oxide film develops some cracks, anodic areas are created on the surface, while the metal parts act as cathode. It follows that the anodic areas are small surface parts, while nearly the rest of the surface of the metal forms large cathodes.

At anodic areas of the metal (Fe) dissolve as ferrous ions with liberation of electrons.

$$Fe \longrightarrow Fe^{+2} + 2e^- \quad \text{(Oxidation)}$$

The liberated electrons flow from anodic to cathodic areas through iron metal, where electrons are intercepted by the dissolved oxygen as,

$$\frac{1}{2}O_2 + H_2O + 2e^- \longrightarrow 2OH^- \quad \text{(Reduction)}$$

The Fe^{+2} ions and OH^- ions diffuse and then they meet where ferrous hydroxide is precipitated.

$$Fe^{+2} + 2OH^- \longrightarrow Fe(OH)_2\downarrow$$

1. If enough oxygen is present, ferrous hydroxide is easily oxidized to ferric hydroxide.

$$4Fe(OH)_2 + O_2 + 2H_2O \longrightarrow 4Fe(OH)_3$$

This product is called yellow rust actually corresponds to $Fe_2O_3 . H_2O$

2. If the supply of oxygen is limited, the corrosion product may be even black anhydrous magnetite, Fe_3O_4.

3.3.1 Galvanic Corrosion

When two dissimilar metals (*e.g.*, Zn and Cu) are electrically connected and exposed to an electrolyte, the metal higher in electrochemical series undergoes corrosion. This type of corrosion is called **Galvanic Corrosion.** In the above example, Zn (higher in electrochemical series) forms the anode and is attacked and gets dissolved; whereas Cu (lower) acts as cathode.

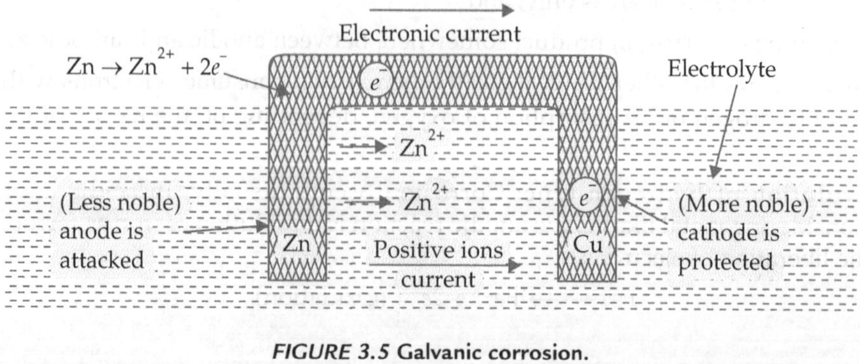

FIGURE 3.5 **Galvanic corrosion.**

Mechanism. In acidic solution, the corrosion occurs by the hydrogen evolution process; while in neutral or slightly alkaline solution, oxygen absorption occurs. The current flows from anode to cathode.

$$Zn \longrightarrow Zn^{+2} + 2e^-$$

Thus, it is evident that the corrosion occurs at the anodic metal; while the cathodic part is protected.

Example:

1. Steel screws in a brass marine hardware,
2. Pb-Sb solder around copper wire.

3.3.2 Concentration Cell Corrosion (Differential Aeration Corrosion)

Corrosion due to electrochemical attack on the metal surface exposed to an electrolyte of varying concentrations of varying aeration is known as **Concentration Cell Corrosion.**

Differential aeration Corrosion occurs when one part of metal is exposed to a different air concentration from the other part. This causes a difference in potential between differently aerated areas.

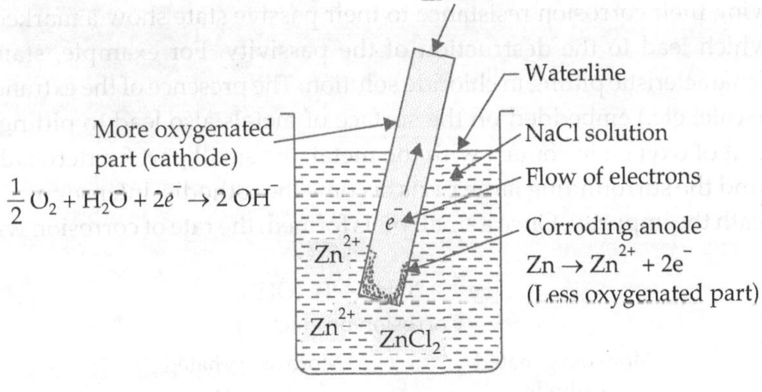

FIGURE 3.6

Differential aeration accounts for the corrosion of metals, partially immersed in a solution, just below the waterline. Thus, if a metal (Zn) is partially immersed in a dilute solution of a neutral salt (NaCl) and the solution is not agitated properly, then the parts above and closely adjacent to the waterline are more strongly aerated (because of the easy access of the oxygen) and hence, become cathodic. On the other hand, parts immersed to greater depth show a smaller oxygen concentration and thus, become anodic. So, a difference of potential is created, which causes a flow of current between the two differentially aerated areas of the same metal. Zn will dissolve at the anodic areas and oxygen will take up electrons at the cathodic areas to form hydroxyl ions.

$$Zn \longrightarrow Zn^{+2} + 2e^- \quad \text{(Oxidation)}$$

$$\frac{1}{2} O_2 + H_2O + 2e^- \longrightarrow 2OH^- \quad \text{(Reduction)}$$

This circuit is completed by migration of ions, through the electrolyte and flow of electrons through the metal from anode to cathode.

In a similar way, iron corrodes under drops of water (or salt solution). Areas covered by droplets, having no access of oxygen, become anodic with respect to the other areas, which are freely exposed to air.

From the above, it is clear that oxygen concentration cell increases corrosion, but it occurs where the oxygen concentration is lower.

3.3.3 Pitting Corrosion

Pitting corrosion is a localized accelerated attack, resulting in the formation of cavities around which the metal is relatively unattached. Thus, pitting corrosion results in the formation of pinholes, pits and cavities in the metal. Pitting is, usually, the result of the breakdown or cracking of the protective film on a metal at specific points. This gives rise to the formation of small anodic and large cathodic areas. In the correct environment, this produces corrosion current. Breakdown of the protective film may be caused by: (*i*) surface roughness or non-uniform finish, (*ii*) scratches or cut edges, (*iii*) local straining of metal, due to non-uniform stresses, (*iv*) alternating stresses, (*v*) sliding under load, (*vi*) impingement attack (caused by turbulent flow of a solution over a metal surface), and (*vii*) chemical attack.

Metals owing their corrosion resistance to their passive state show a marked pitting under all conditions, which lead to the destruction of the passivity. For example, stainless steel and aluminum show characteristic pitting in chloride solution. The presence of the extraneous impurities (like sand, dust, scale, etc.) embedded on the surface of metals also lead to pitting. Owing to the differential amount of oxygen in contact with the metal, the small part (underneath the impurity) becomes anodic and the surrounding large parts become the cathodic. Intense corrosion, therefore, start just underneath the impurity. Once a small pit is formed, the rate of corrosion will be increased.

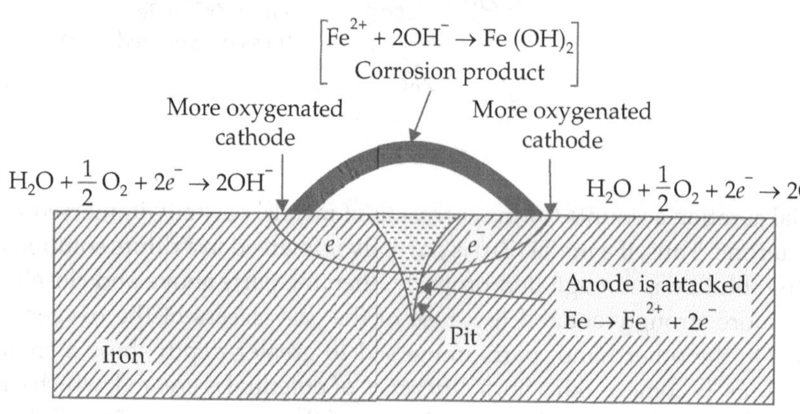

FIGURE 3.7 **Pitting corrosion.**

3.3.4 Waterline Corrosion

When water is stored in a steel tank, it is generally found that the maximum amount of corrosion takes place along a line just beneath the level of the water meniscus. The area above the waterline (highly-oxygenated) acts as the cathodic and is completely unaffected by corrosion. However, if the water is relatively free from acidity, little corrosion takes place.

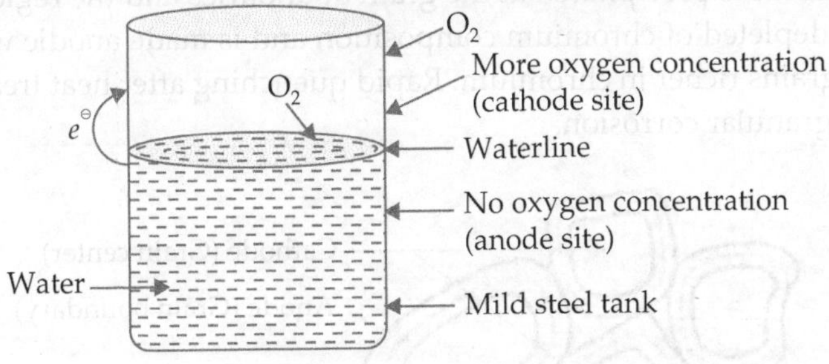

FIGURE 3.8 **Waterline corrosion.**

The problem of waterline corrosion is also that concerns marine engineers. In the case of ships, this kind of corrosion is often accelerated by marine plants attaching themselves to the sides of ships. The uses of special antifouling paints restrict this to some extent.

3.3.5 Stress Corrosion

Stress corrosion (or stress cracking) is the combined effect of static tensile stresses and the corrosive environment on a metal. It is characterized by a highly localized attack occurring, when overall corrosion is negligible.

For stress corrosion to occur: (*i*) presence of tensile stress, and (*ii*) a specific corrosive environment are necessary. The corrosive agents are highly specific and selective such as (*a*) caustic alkalis and strong nitrate solution for mild steel; (*b*) traces of ammonia for brass; (*c*) acid chloride solution for stainless steel.

This type of corrosion is seen in fabricated articles of certain alloys (like high-zinc brass and nickel brass) due to presence of stress caused by heavy working like rolling, drawing or insufficient annealing. However, pure metals are relatively immune of stress corrosion.

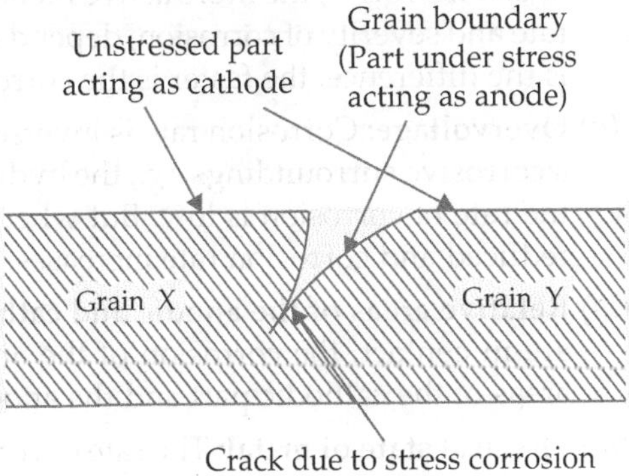

FIGURE 3.9 **Stress corrosion.**

Stress corrosion involves in a localized electrochemical corrosion, occurring along narrow paths, forming anodic areas with respect to the more cathodic areas at the metal surface. Presence of stress produce strains, which result in localized zones of higher electrode potential. These become so chemically-active that they are attacked, even by a milk corrosive environment, resulting in the formation of a crack, which grows and propagates in a plant (perpendicular to the operating tensile stress), until failure occurs or it may stop, after progressing a finite distance.

Intergranular corrosion: This corrosion is observed in case of alloys. The corrosion product is observed at the boundaries of grains. Externally, it is not seen. There is a sudden failure of material due to this corrosion. For example, during the welding of stainless steel (an alloy of Fe, C, Cr), chromium carbide is precipitated at the grain boundaries and the region adjacent to grain boundaries become depleted of chromium composition and is made anodic with respect to solid solution within the grains richer in chromium. Rapid quenching after heat treatment of a metal is the remedy of inter-granular corrosion.

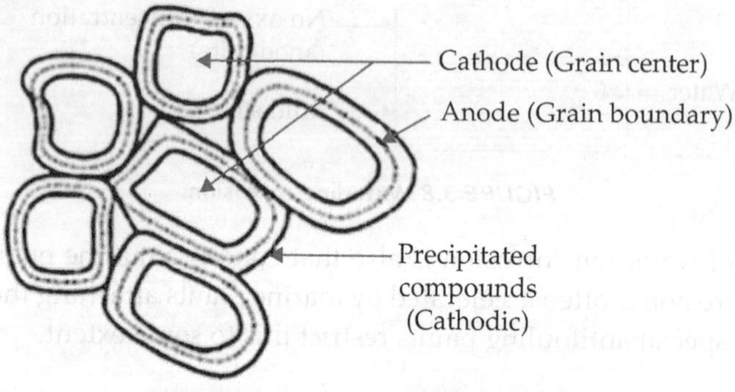

FIGURE 3.10 **Intergranular corrosion.**

3.4 FACTORS INFLUENCING CORROSION

The rate and extent of corrosion, depends on the following factors:

1. Nature of the Metal

(*i*) **Position in galvanic series:** When two metals or alloys are in electrical contact, in presence of an electrolyte, the more active metal (or higher up in the series) suffers corrosion. The rate and severity of corrosion, depends upon the difference in their positions, and greater is the difference, the faster is the corrosion of the anodic metal/alloy.

(*ii*) **Overvoltage:** Corrosion rate is inversely proportional to the overvoltage of the metal in a corrosive surroundings *e.g.*, the hydrogen overvoltage of Zn in 1 M H_2SO_4 is 0.70 V. So the rate of corrosion is low. But when some Cu impurity is present, the overvoltage is reduced and corrosion rate increases.

(*iii*) **Relative areas of the anodic and cathodic parts:** When two dissimilar metals or alloys are in contact, the corrosion of the anodic part is directly proportional to the ratio of areas of the cathodic part and the anodic part.

(*iv*) **Physical state of metal:** The rate of corrosion is influenced by physical state of the metal (such as grain size, orientation of crystals, stress, etc.). The smaller the grain-size of the metal or alloy, the greater will be its solubility and hence, greater will be its corrosion. Moreover, areas under stress, even in a pure metal, tend to be anodic and corrosion takes place at these areas.

(*v*) **Nature of surface film:** In aerated atmosphere, practically all metals get covered with a thin surface film (thickness = a few Angstroms) of metal oxide. The ratio of the volumes of the metal oxide to the metal is known as a 'specific volume ratio'. Greater the specific volume ratio, lesser is the oxidation corrosion rate. The specific volume ratios of Ni, Cr, and W are 1.6, 2.0 and 3.6 respectively. Consequently, the rate of oxidation of tungsten is least, even at elevated temperatures.

2. Nature of the Corroding Environment

(*i*) **Temperature:** With increase of temperature of environment, the reaction as well as diffusion rate increase, thereby corrosion rate is generally enhanced.

(*ii*) **Humidity of air:** Humidity of air is the deciding factor in atmospheric corrosion. "Critical humidity" is defined as the relative humidity above which the atmospheric corrosion rate of metal increases sharply. The value of critical humidity depends on the physical characteristics of the metal as well as nature of the corrosion products. The reason why corrosion of a metal becomes faster in humid atmosphere is that gases (CO_2, O_2 etc) and vapours, present in atmosphere furnish water to the electrolyte, essential for setting up an electrochemical corrosion cell.

(*iii*) **Presence of impurities in a atmosphere:** Atmosphere, in the vicinity of industrial areas, contains corrosive gases like CO_2, H_2S, SO_2, and fumes of HCl, H_2SO_4 etc. In presence of these gases, the acidity of the liquid, adjacent to the metal surfaces, increases and its electrical conductivity also increases. This consequently, results in an increase of corrosion current flowing in the local electrochemical cells on the exposed metal surfaces.

(*iv*) **Influence of pH:** Generally, acidic media (*i.e.*, pH < 7) are more corrosive than alkaline and neutral media. However, amphoteric metals (like Al, Zn, Pb, etc.) dissolve in an alkaline solution as complex ions.

(*v*) **Conductance of the corroding medium:** In corrosion of underground or submerged structures, the conductance of the medium is of prime importance. Conductance of dry sandy-soils is lower than those of clayey and mineralized soils. Consequently, stray currents (from power leakages) will cause more severe damage to the metallic structures, buried under clayey and mineralized soils than those under dry sandy-soils.

(*vi*) **Formation of oxygen concentration cell:** With the increase in supply of oxygen/air to the moist-metal surface, the corrosion is promoted. Less oxygen concentration (*e.g.* oxide-coated part or less-exposed part) parts becomes anodic; while the more-oxygenated regions (or parts more exposed to oxygen) becomes cathodic, thereby leading to the formation of "oxygen-concentration cell"in which the anodic part suffers corrosion.

3.5 CORROSION CONTROL METHODS (PROTECTIVE COATINGS)

3.5.1 Cathodic Protection

The principle involved in this method is to force the metal to be protected to behave like a cathode, thereby corrosion does not occur. There are two types of cathodic protections:

(*i*) **Sacrificial anodic protection method:** In this protection method, the metallic structure (to be protected) is connected by a wire to a more anodic metal, so that all the corrosion is concentrated at this more active metal. The more active metal itself gets corroded slowly; while the parent structure (cathodic) is protected. The more active metal so-employed is called "sacrificial anode". The corroded sacrificial anode block is replaced by a fresh one, when consumed completely. Metals commonly employed as sacrificial anodes are magnesium, zinc, aluminum and their alloys. Important applications of sacrificial anodic method include protection of buried pipelines, underground cables; marine structure, ship-hulls, water-tanks, piers, etc.

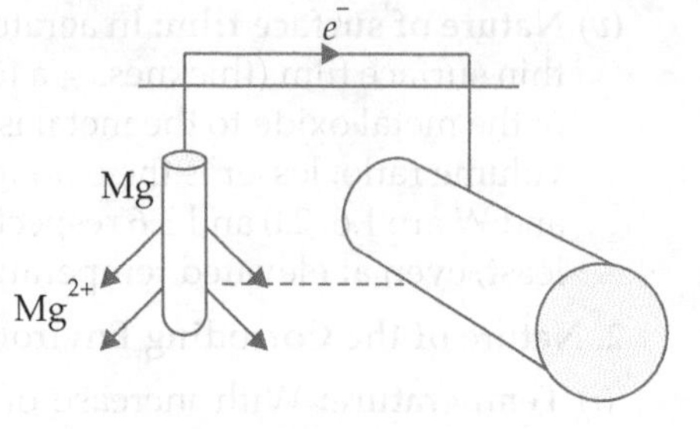

FIGURE 3.11

(*ii*) **Impressed current cathodic protection:** In this method, an impressed current is applied in opposite direction to nullify the corrosion current, and convert the corroding metal from anode to cathode. Usually, the impressed current is derived from a direct current source (like battery or rectifier on A.C. line) with an insoluble anode (like graphite, high silica iron, scrap iron, stainless steel or platinum). Usually, a sufficient D.C. current is applied to an insoluble anode, buried in the soil (or immersed in the corroding medium), and connected to the metallic structure to be protected. The anode is, usually, in a backfill (composed of coke breeze or gypsum) so as to increase the electrical contact with the surrounding soil. This type of cathodic protection has been applied to open water-box coolers, water-tanks, buried oil or water pipes, condensers, transmission line towers, marine piers, laid-up ships, etc. This kind of protection technique is, particularly, useful for large structures for long-term operations.

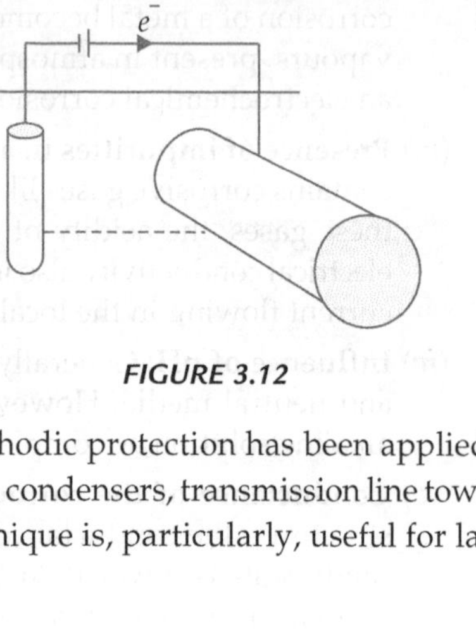

FIGURE 3.12

3.6 APPLICATION OF PROTECTIVE COATINGS

Protecting the surface of an object by the application of coating is probably the oldest of the common procedures for corrosion prevention. A coated surface isolates the underlying metal from the corroding environment. The only limitations of this method are the service behaviour of the protective coatings: (*i*) the coating applied must be chemically inert to the environment under particular conditions of temperature and pressure, (*ii*) Moreover, coatings must prevent the penetration of the environment to the material, which they protect. A brief description of two important protective coatings is given below.

1. Anodic coatings   2. Cathodic coatings.

3.6.1 Anodic Coatings

Anodic coatings are produced from coating-metals, which are "anodic" to the base metal (*i.e.*, which is to be protected). For example, coating of Zn, Al and Cd on steel are anodic, because their electrode potentials are lower than that of the base metal, iron. If any pores, breaks or discontinuities occur in such an anodic coating, a galvanic cell is formed between the coating-metal and the exposed part of base metal, *e.g.*, in case of galvanized steel, zinc, the coating-metal (being anodic) is attacked; leaving the underlying cathodic metal (iron) unattacked. Galvanic cell is formed between Zn and the exposed iron. Zinc (being anodic to iron) dissolves anodically; whereas the iron (being cathodic) is protected. Thus, no attack on the iron occurs, until practically all the zinc has first corroded in the vicinity of the exposed iron spot. So, zinc coating protect iron sacrificially.

3.6.2 Cathodic Coatings

Cathodic coatings are obtained by coating a more noble metal (having higher electrode potential) than the base metal. They protect the base metal, because they have higher corrosion-resistance than the base metal. Cathodic coating provides effective protection to the base metal only when they are completely continuous and free from pores, breaks or discontinuities.

If such coatings are punctured, much more corrosion damages can be done to the base metal than to metal without it. For example, a tin-coating on a sheet of iron provides protection only as long as the surface of the metal is completely covered, since tin is lower than iron electromotive series. However, if the surface coating is punctured, the tin becomes the cathode; while the exposed iron (which above tin in the electromotive series) acts as anode. A galvanic cell is set up and an intense localized attack at the small exposed part occurs, resulting in severe pitting and perforation of the base metal, iron. Such combination of a small anode and large cathode area is always very dangerous.

Anodic versus cathodic coatings

Anodic coatings	*Cathodic coatings*
1. Protects the underlying base metal 'sacrificially'.	Protects the underlying base metal, due to its noble character and higher corrosion-resistance.
2. Electrode potential of coating metal is lower than that of the base metal.	Electrode potential of coating metal is higher than that of the base metal.
3. If pores, breaks or discontinuities occur in such a coating, the base metal is not corroded, till all the coating metal is consumed.	If pores, breaks or discontinuities occur in such a coating the, corrosion of the base metal is speeded up.
4. Coating of Zn on iron is an example.	Coating of Sn on iron is an example.

3.7 METHODS OF APPLICATION OF METAL COATINGS

3.7.1 Hot Dipping

Hot dipping is used for producing a coating of low-metal such as Zn (m.p. = 419°C), Sn (m.p. = 232°C), Pb, Al, etc. on iron, steel and copper, which have relatively higher melting points. The process, ingeneral, consists of immersing the base metal in a bath of the molten coating-metal, covered by a molten flux layer (usually zinc chloride). The flux cleans the base metal surface and prevents the oxidation of the molten coating-metal. For good adhesion, the base metal surface must be very clean; otherwise it cannot be properly wetted by the molten metal. Two most widely applied hot dipping methods are:

(*a*) **Galvanizing:** Galvanizing is the process of coating iron or steel sheets with a thin coat of zinc to prevent them from rusting. The iron or steel article (*e.g.*, sheet, pipe, and wire) is first cleaned by pickling with dilute H_2SO_4 solution for 15–20 minutes at 60–90°C. This treatment also removes any scale, rust (oxide layer) and impurities. The article is then washed well and dried. It is then dipped in bath of molten zinc, maintained at 425–430°C. The surface of the bath is kept covered with a flux [NH_4Cl] to prevent oxide formation. When the article is taken out, it is found to have been coated with a thin layer of zinc. It is then passed through a pair of hot rollers. This process removes any surplus of zinc and produces a thin film of uniform thickness. Then, it is annealed at a temperature of 650°C and finally, cooled slowly.

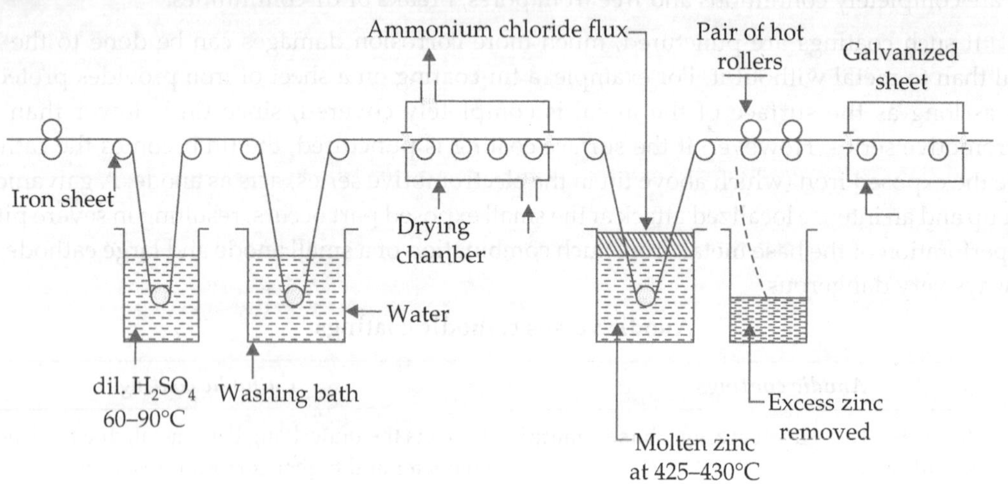

FIGURE 3.13 **Galvanizing.**

Uses. It is most widely used for protection of iron from atmospheric corrosion in the form of roofing sheets, wires, pipes, nails, bolts, screws, buckets, tubes, etc. It may be pointed here that zinc gets dissolved in dilute acids to form highly toxic (or poisonous) compounds. Hence, galvanized utensils cannot be used for preparing and storing foodstuffs, especially acidic ones.

(*b*) **Tinning:** Tinning is coating tin over the iron or steel articles. The process consists in first treating steel sheet in dilute sulphuric acid (pickling) to remove any oxide film. After this, it is passed through a bath of zinc chloride flux. The flux helps the molten metal to adhere to the metal sheet. Next, the sheet passes through a tank of molten tin and finally through a series of rollers from underneath the surface of a layer palm oil. The palm oil protects the hot tin-coated surface against oxidation. The rollers remove any excess of tin and produce a thin film of uniform thickness on the steel sheet.

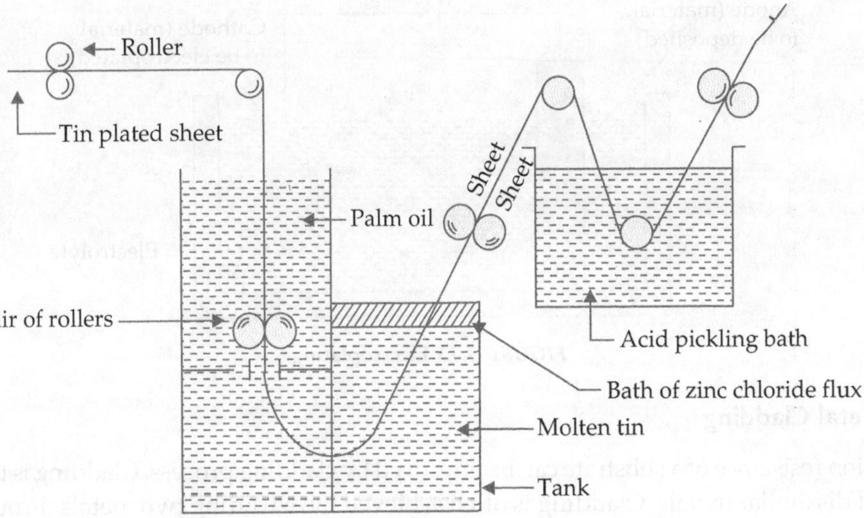

FIGURE 3.14 **Tinning.**

Uses. Tin possess considerable resistance against atmospheric corrosion. Moreover, because of non-toxic nature of tin, tinning is widely used for coating steel, copper and brass sheets, used for manufacturing containers for storing foodstuffs, ghee, oils, kerosene and packing food materials. Tinned-copper sheets are employed for making cooking utensils and refrigeration equipments.

3.7.2 Electroplating

The process of coating base metals with protective metals like Cu, Ni, Zn, Pb, Sn, Au and Ag etc., is known as **Electroplating.**

Process: Before electroplating the metal surface is cleaned thoroughly. The article to be electroplated is made cathode since metallic ions are positive and thus get deposited on the cathode. The anode is made of pure metal, which is to be coated on the article. The electrolyte is the salt of the metal to be coated on the article. A direct current (D.C.) is passed through the electrolyte. The anode dissolves, depositing the metal ions from the solution on the article in the form of a metallic coating.

Electroplating with Gold

Cathode: Article to be electroplated (spoon).

Anode: Block of gold metal.

Electrolyte: Aqueous solution of $AuCl_3$ or potassium auro-cyanide $K[Au(CN)_2]$.

Factors effecting electroplating: Cleaning of article, Electrolyte concentration, Selection of electrolyte, pH of electrolyte, thickness of the deposit and additives. These are all the factors that influence electroplating.

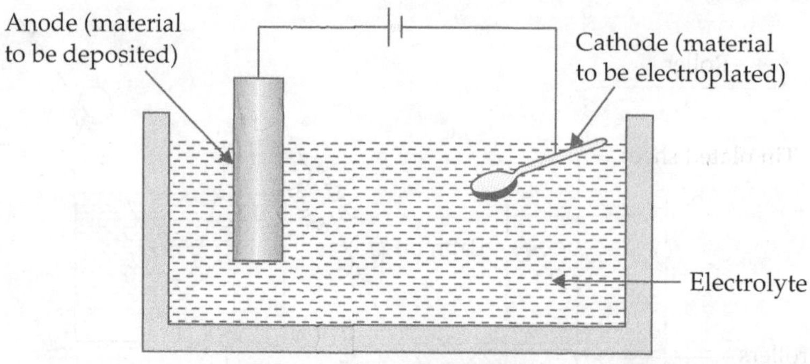

FIGURE 3.15 **Electroplating.**

3.7.3 Metal Cladding

The corrosion resistance of a substrate can be improved by cladding process. Cladding is the bonding together of dissimilar metals. Cladding is often achieved by extruding two metals through a die as well as pressing or rolling sheets together under high pressure. Cladding is done by the following process:

In this process a dense, homogeneous layer of coating metal is bonded firmly and permanently to the base metal on one or both sides. This is known as metal cladding. The choice of cladding material depends on the corrosion resistance required for any particular environment.

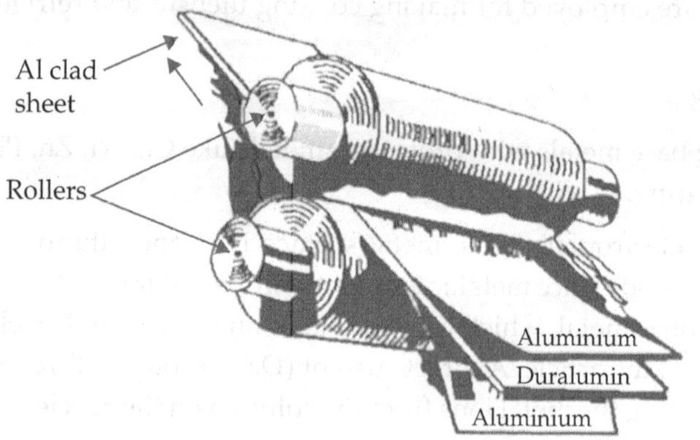

FIGURE 3.16 **Metal cladding.**

The corrosion resistant metals like Ni, Cu, Pb, Ag, Pt etc., and alloys like Stainless steel, Nickel alloys, Copper alloys, Lead alloys are used as cladding materials. The common base metals cladded are mild steel, aluminium, copper, nickel and their alloys.

Generally cladding is done by arranging thin sheets of the coating metal and the base metal sheet in the form a sandwich, which is then passed through rollers, under the action of heat and pressure. A plate of duralumin is sandwiched between two layers of 99.5% pure aluminium produce 'alclad' sheeting, which is widely used in aircraft industry.

3.7.4 Electroless Plating (Chemical Deposition Process)

A technique of depositing a noble metal (from its salt solution) on a catalytically active surface of a less noble metal by employing a suitable reducing agent without using electric current is known as **Electroless plating**.

Or

Electroless plating is similar in result to electroplating in that a metallic layer is deposited onto the surface of a part. Electro less plating, however, uses a chemical deposition process-instead of an external electrical current-to achieve the desired result.

- Nickel is the most common plating material deposited in electroless plating processes.
- Electroless plating provides a more even plate layer than electroplating, even on complex geometries.

In this process added reducing agent causes the reduction of the metallic ions to metal, which eventually gets plated over the catalytically activated surface giving a highly uniform, but thin coating. Thus:

$$\text{Metal ions} + \text{Reducing agent} \longrightarrow \text{Metal} + \text{Oxidised product}$$

Various aspects of electroless plating are:

1. Preparation of active surface of the object to be plated:

This step is most important in this technique. This is achieved by using one of the following methods:

(*i*) Etching, *i.e.*, by acid treatment.

(*ii*) Electroplating a thin layer of the metal to be plated or any other suitable metal. This is usually followed by heat treatment.

(*iii*) Treating with $SnCl_2$, followed by dipping in $PdCl_2$ solution. This treatment yields a thin layer of Pd on the treated surface. Usually; this is followed in case of plastics and printed circuit boards.

2. Plating bath is composed of:

(*i*) Soluble salt of metal (like chloride or sulphate) to be plated.

(*ii*) Reducing agent like formaldehyde, hypophosphite, etc.

(*iii*) Complexing agent like tartrate, citrate, succinate, etc. This improves the quality of the deposit.

(*iv*) Exaltant like succinate, fluoride, glycinate, etc. This increases the plating rate.

(*v*) Stabilizer like thiourea, cations of Pb, Ca, Tallium, etc., is added to prevent the composition of the plating bath solution.

(*vi*) Buffer like sodium stearate, sodium hydroxide + Rochelle salt, etc., is added to control the pH of the bath. It may be pointed that pH of the bath plays a "critical role" in getting a uniform, but thin deposit.

Advantages of electroless plating over electroplating:

1. No electrical energy is required.
2. Better throwing power.

3. Plating on articles made of insulators (like plastics) and semiconductors can easily be carried out.
4. Even intricate parts (of irregular shapes) can be plated uniformly.
5. Electroless plated coating process unique mechanical, chemical and magnetic features.

3.8 PAINTS

Paint is a mechanical dispersion mixture of one or more pigments in a vehicle. The 'vehicle' is a liquid, consisting of non-volatile, film-forming material, drying oil and a highly volatile solvent, thinner. When a paint is applied to a metal surface (usually by brushing or spraying), the thinner evaporates; while the drying oil slowly oxidizes forming a dry pigmented-film.

Requisites of a good paint. A good paint possesses the following characteristics:

1. It should be fluid enough to be spread easily over the protected surface.
2. It should possess high covering power.
3. It should form a quite tough, uniform, adherent and impervious film.
4. Its film should not get cracked on drying.
5. It should protect the painted surface from corrosion effects of environment.
6. It should form film, the colour of which is quite stable to the effect of atmosphere and other agencies.
7. Its film should be glossy (*i.e.*, having shine or luster).
8. Its film should be stable.
9. It can be prepared in such a consistency as to be easily applicable with brush or spraying device and that it yields a smooth and uniform surface.
10. It should possess high adhesion capacity to the material over which it is intended to be used.

3.8.1 Constituents of Paints

1. Pigment. Pigment is a solid substance, which is an essential constituent of paint. Its functions are to: (*i*) provide capacity to paint, (*ii*) provide strength to paint, (*iii*) provide desired colour to paint, (*iv*) give aesthetical appeal (*i.e.*, pleasing to look at) to the paint film, (*v*) give protection to the paint film by reflecting harmful ultraviolet light, (*vi*) provide resistance to paint film against abrasion/wear, (*vii*) Improve the impermeability of paint film to moisture, and (*viii*) increase wheather-resistance of the film.

Pigments used should whitele [such as white lead, zinc oxide, lithophone, titanium oxide] or coloured [(*i*) red-red lead, ferric oxide, Venetian red, chrome red, (*ii*) green-chromium oxide, (*iii*) blue-prussian blue, (*iv*) black-carbon black].

Characteristics of good pigments: Good pigments should be: (*i*) opaque, (*ii*) chemically inert so that paint film has stability and longer life, (*iii*) non-toxic so that there is no bad effect on the health of painter as well as inhabitants, (*iv*) freely mixable with film-forming constituent, oil, and (*v*) cheap.

2. Vehicle or drying oil. Vehicle or drying oil is a film-forming constituent of the paint. These are glyceryl esters of high molecular-weight fatty acids, generally, present in animal and vegetable oils.

$$
\begin{array}{l}
CH_2OCOR \\
| \\
CHOCOR \\
| \\
CH_2OCOR
\end{array}
$$

The most widely used drying oils are linseed oil, soya bean oil, and dehydrated castor oil.

Functions of drying oils: Drying oil supplies to paint-film: (*i*) main film-forming constituent, (*ii*) vehicle or medium, (*iii*) toughness, (*iv*) adhesion, (*v*) durability, and (*vi*) water-proofness.

Reactions in drying of oils: The oil film, after it has been applied on the protected surface, absorbs oxygen (of the air) at the double bonds, forming peroxides, diperoxides and hydro-peroxides, which isomerise, polymerize and condense to form a characteristic tough, coherent, hard, elastic, insoluble, infusible, highly cross-linked structured macromolecular film. The final hardened product actually resembles a thermosetting resin in chemical structure.

3. Thinners

(*i*) Reduce the viscosity of the paint to suitable consistency, so that it can easily be handled and applied;

(*ii*) Dissolve vehicle and the additives in the vehicle;

(*iii*) Suspend the pigments;

(*iv*) Increase the penetration power of the vehicle;

(*v*) Increase the elasticity of the paint film;

(*vi*) Help the drying of the paint film, as they evaporate.

Common thinners used are turpentine, mineral spirits (from petroleum), benzene, di-pentene, naphtha, xylol, kerosene, methylated naphthalene, etc.

4. Driers. Driers are oxygen-carrier catalysts. They accelerate the drying of the oil-film through oxidation polymerization and condensation. Thus, their main function is to improve the drying quality of the oil-film. The most effective driers are resonates, linoleates, tungstates and naphthenates of Co, Mn, Pb and Zn.

5. Extenders or fillers. Extenders or fillers are low refractive indices materials, generally, of white colour, added to: (*i*) reduced the cost, (*ii*) increase durability of the paint, (*iii*) provide negligible covering power to the paint, (*iv*) help to reduce the cracking of dry paint film and sometimes help to keep the pigments in suspension, (*v*) serve to fill voids in the film, (*vi*) increase random arrangement of pigment of pigment particles, and (*vii*) act as carrier for the pigment colour.

Important extenders used are barites ($BaSO_4$), talc, asbestos, ground silica, gypsum, ground mica, state powder, china-clay, whitting ($CaCO_3$), magnesium silicate, diamite clay, calcium sulphate, etc.

6. Plasticizers. Sometimes plasticizers are incorporated in the paint: (*i*) to provide elasticity to the film, and also (*ii*) to minimize its cracking. Common plasticizers used are tri-cresyl phosphate, tri-phenyl phosphate, tri-butyl phthalate, di-amyl phthalate and di-butyl tatrate.

7. Anti-skinning agents. Anti-skinning agents are, sometimes, added to some paints with the object preventing gelling and skinning of the paint film. Important anti-skinning agents are polyhydroxy phenols.

EXERCISE

1. Define Corrosion. Mention the causes and effects of corrosion.
2. Define chemical corrosion. Explain about the following:

   (*a*) Oxidation corrosion (*b*) Corrosion by other gases

   (*c*) Liquid metal corrosion.
3. Define electrochemical corrosion. Explain about the following:

   (*a*) Galvanic corrosion

   (*b*) Concentration cell corrosion (differential aeration corrosion)

   (*c*) Pitting corrosion

   (*d*) Waterline corrosion

   (*e*) Stress corrosion

   (*f*) Intergranular corrosion.
4. What are the factors that effecting corrosion?
5. Explain about the following cathode protection process:

   (*a*) Sacrificial anodic process (*b*) Impressed current method.
6. Explain about the following methods of surface coatings on metals:

   (*a*) Hot dipping-Galvanizing,Tinning (*b*) Metal cladding

   (*c*) Electroplating (*d*) Electroless plating.
7. What is paint? Explain their constituents and their functions.

Chapter **4**

HIGH POLYMERS

4.1 INTRODUCTION

Polymer. In Greek poly = many, mers = units or parts

Small molecules combine with each other to form big molecules. The big molecules are called polymers. Polymers are macro molecules or giant molecules of high molecular masses.

Example. Polyethylene

$$\underset{\text{Ethylene}}{nCH_2{=}CH_2} \longrightarrow \underset{\text{Polyethylene}}{\left[CH_2{-}CH_2 \right]_n}$$

Polymerization. The phenomenon of the formation of the polymers is known as polymerization.

Example. Polyethylene

$$\underset{\text{Ethylene}}{nCH_2{=}CH_2} \longrightarrow \underset{\text{Polyethylene}}{\left[CH_2{-}CH_2 \right]_n}$$

Monomer. Reactant molecules in the formation of polymers are called **monomers.**

Example. Ethylene is the monomer in the formation of polyethylene.

$$\underset{\text{Ethylene (Monomer)}}{nCH_2{=}CH_2} \longrightarrow \underset{\text{Polyethylene}}{\left[CH_2{-}CH_2 \right]_n}$$

Repeating unit. The unit which is repeated in the polymer is known as **repeating unit**.

Example. Ethylene

$$\underset{\text{Ethylene}}{nCH_2{=}CH_2} \longrightarrow \underset{\text{Repeating unit}}{\left[CH_2{-}CH_2{-}CH_2{-}CH_2{-}CH_2{-}CH_2 \right]_n}$$

Degree of polymerization. The number of repeating units in the formation of polymers is known as degree of polymerization. It is denoted by Dp.

$$Dp = \frac{\text{Total molecular weight of the polymer}}{\text{Molecular weight of monomeric unit}}$$

Example. If 100 molecules of ethylene on polymerization give polymeric chain, the degree of polymerization of the polymer is 100.

Functionality. In polymerization, the monomers are linked to each other through covalent bonds. So the monomer should have at least two reacting sites (or) bonding sites.

The number of active bonding sites of a monomer is known as functionality.

1. Bifunctional monomers. Bifunctional monomers contain two reactive sites.

$CH_2{=}CH_2$ Ethylene

$CH_2{=}CH{-}Cl$ Vinyl chloride

$CH_2{=}CH{-}CN$ Acrylonitrile

2. Trifunctional monomers. Trifunctional monomers contain three reactive sites.

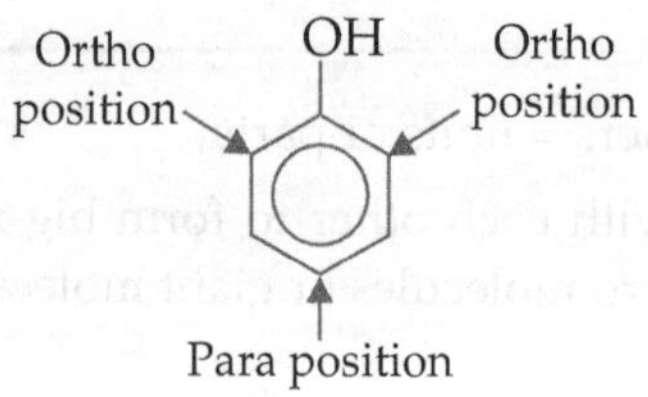

Phenol (2 ortho positions and 1 para position)

3. Poly functional monomers. A poly functional monomer contains more than three reactive sites. If the numbers of reactive sites are more than 3 in a monomer molecule, the three dimensional network polymer is produced.

Opening of one double bond gives 2 reactive sites; opening of another double bond gives 2 more reactive sites.

$$H{-}C{\equiv}C{-}H$$

4.2 TYPES OF POLYMERIZATION

The polymerization is of four types:

1. Addition polymerization or chain growth polymerization.
2. Condensation polymerization or step growth polymerization.
3. Copolymerization.
4. Coordination polymerization or Ziegler-Natta polymerization.

4.2.1 Addition or Chain Polymerization

In polymerization, the polymers are formed without the loss of any material, and the product is an exact multiple of the original monomeric molecule such type of polymerization is known as Addition polymerization. The addition polymerization reaction must be instigated by the application of heat, light, pressure or a catalyst for breaking down the double bonds of monomers.

Example:

1. Polyethylene

$$\underset{\text{Ethylene}}{CH_2{=}CH_2} \longrightarrow \underset{\text{Polyethylene}}{\left[\!-CH_2{-}CH_2-\!\right]_n}$$

2. Polyvinyl chloride (PVC)

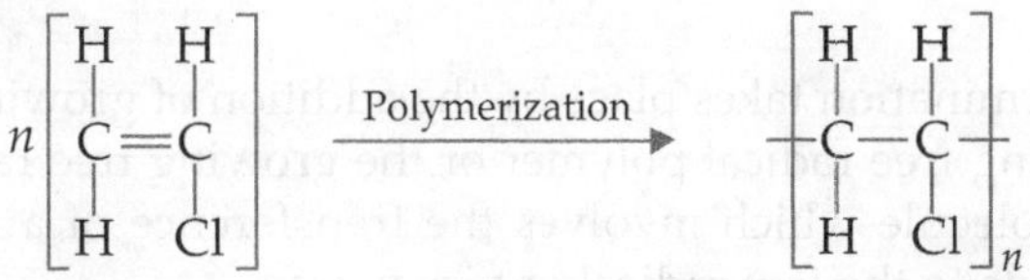

Vinyl chloride Polyvinyl chloride

3. Polystyrene

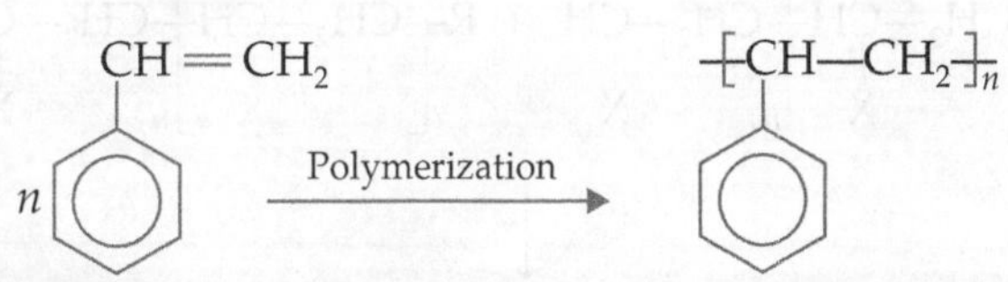

Styrene Polystyrene (PS)

Mechanism of Addition or Chain Polymerization. Addition polymerization reactions proceed by a chain reaction mechanism consisting of three important steps,

(*i*) Initiation or the formation of an active centre,

(*ii*) Propagation or the formation of a polymer having the active centre, and

(*iii*) Termination or removal of the active centre.

The three different active centres found to be formed during the addition polymerization. They are Free radical, Carbonium ion, and Carbanion. So the mechanisms involving these reactive species in polymerization constitute the mechanism of addition polymerization. Let us take one by one.

(*a*) **Free radical addition polymerization.** It involves three steps, these steps will form only linear molecules.

(*i*) **Initiation.** A reagent that is either itself a free radical, such as oxygen molecule, or that easily decomposes to form free radicals, such as organic peroxides, is used to initiate the reaction.

$$C_6H_5CO{-}O{-}O{-}CO{-}C_6H_5 \longrightarrow 2C_6H_5COO^*$$

$$2C_6H_5COO^* \longrightarrow 2C_6H_5^* + CO_2$$

$$R{-}O{-}O{-}R \longrightarrow 2R^* + O_2$$

The radical now adds to the monomer to form a new free radical, *e.g.*, it adds on vinyl chloride in the following way.

$$R^* + CH_2{=}\underset{\substack{|\\X}}{CH} \longrightarrow R{-}CH_2{-}\underset{\substack{|\\X}}{\overset{*}{C}H}$$

(*ii*) **Propagation.** The new free radical now adds to another molecule of monomer to form another new free radical leads to the formation of large free radical.

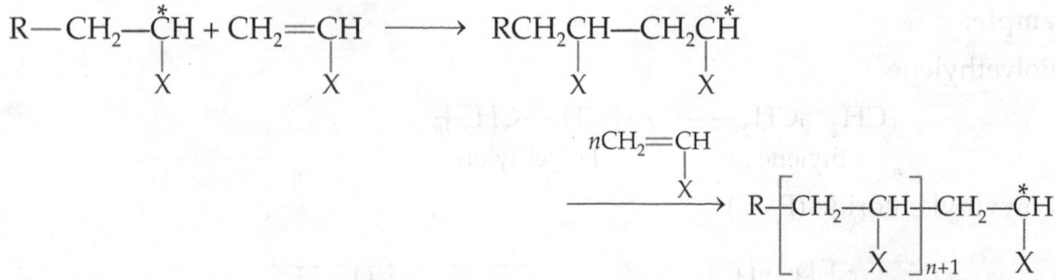

(*iii*) **Termination.** Termination takes place by the addition of growing free radical polymer with other growing free radical polymer or the growing free radical chain reacts with the monomer molecule which involves the transference of a proton either from the monomer molecule to the free radical or vice versa.

Combination:

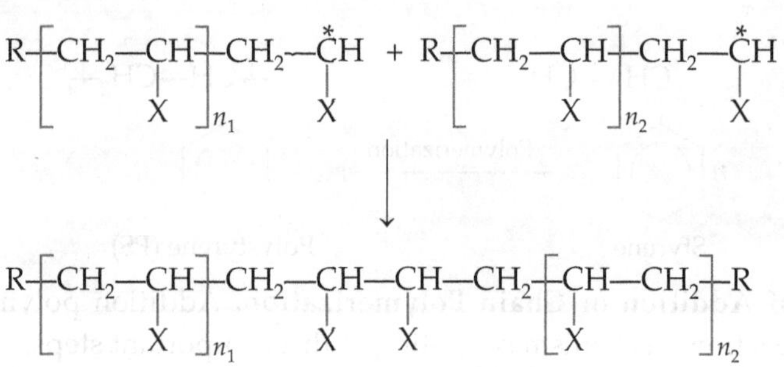

Disproportionation:

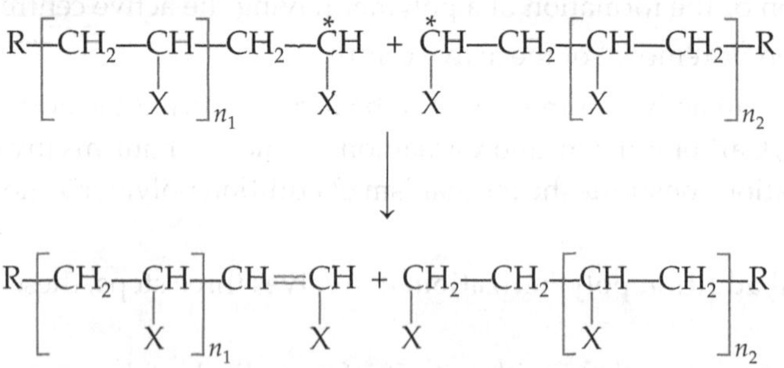

(*b*) **Cationic addition polymerization.** In cationic addition polymerization the chain carries are positively charged carbonium ions. This type of polymerization is catalysed by Friedel Craft's reagents like $AlCl_3$, $TiCl_4$, BF_3, $SnCl_4$ and other strong Lewis acids. It involves three steps:

(*i*) **Initiation step:**

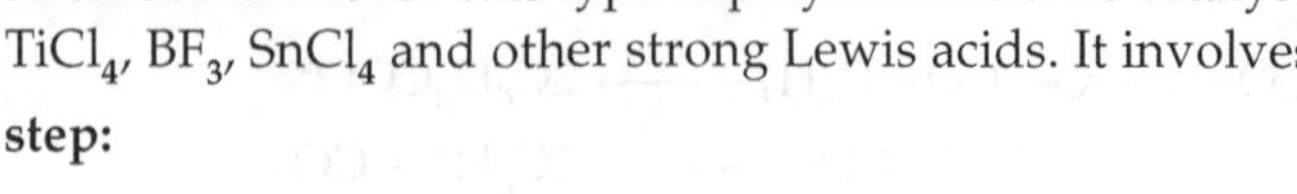

(*ii*) **Propagation step:**

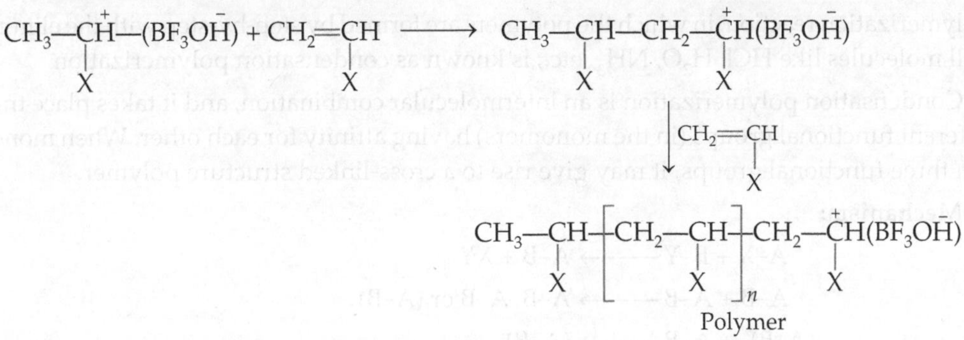

(*iii*) **Termination step:**

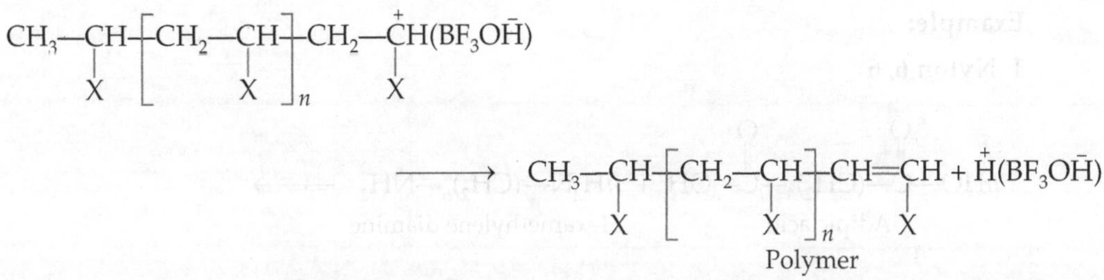

(*c*) **Anionic addition polymerization.** In anionic addition polymerization the chain carriers are negatively charged carbanions. In anionic addition polymerization typical catalysts are strong base, alkali metal amides, and Grignard reagents.

(*i*) **Initiation step.** Carbanion is formed by the addition of monomer with alkali metal amide (KNH_2)

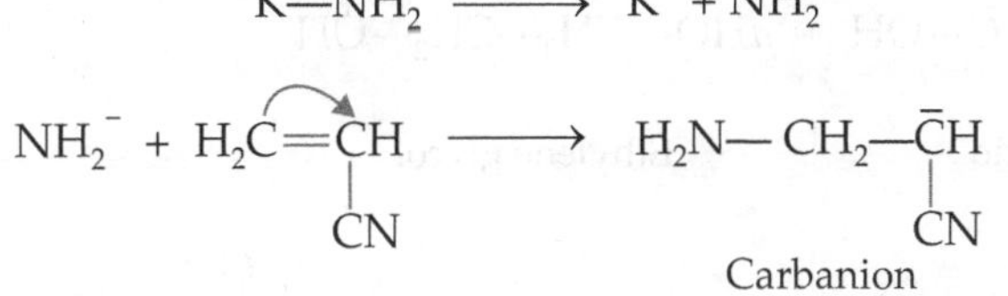

(*ii*) **Propagation step.** Carbanion is propagated by the addition of monomeric molecules to the Carbanion which is formed in the first step.

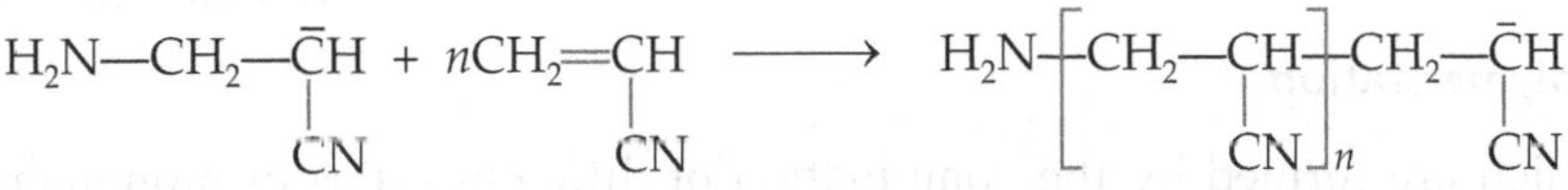

(*iii*) **Termination step.** Reaction is terminated by the addition of ammonia to the carbanion, which is formed in the second step with the liberation of anion.

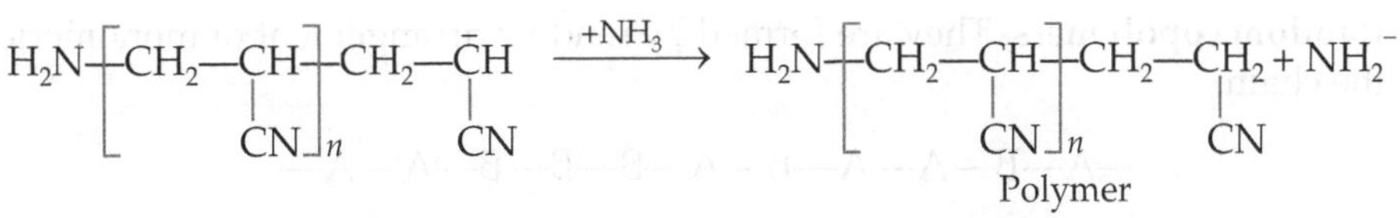

4.2.2 Condensation or Step Polymerization

The polymerization reaction, in which the polymers are formed by step-by-step with the elimination of small molecules like HCl, H_2O, NH_3, etc., is known as condensation polymerization.

Condensation polymerization is an intermolecular combination, and it takes place through the different functional groups (in the monomers) having affinity for each other. When monomers contain three functional groups, it may give rise to a cross-linked structure polymer.

Mechanism:

$$\text{A–X} + \text{B–Y} \longrightarrow \text{A–B} + \text{XY}$$
$$\text{A–B} + \text{A–B} \longrightarrow \text{A–B–A–B or (A–B)}_2$$
$$\text{(A–B)}_2 + \text{A–B} \longrightarrow \text{(A–B)}_3$$
$$\text{(A–B)}_2 + \text{(A–B)}_2 \longrightarrow \text{(A–B)}_4 \longrightarrow \text{(A–B)}_n$$

Example:

1. Nylon 6, 6

$$n\text{HO—}\overset{\text{O}}{\overset{\|}{\text{C}}}\text{—(CH}_2)_4\text{—}\overset{\text{O}}{\overset{\|}{\text{C}}}\text{—OH} + n\text{H}_2\text{N—(CH}_2)_6\text{—NH}_2 \longrightarrow$$

Adipic acid — Hexamethylene diamine

$$\left[\text{—}\overset{\text{O}}{\overset{\|}{\text{C}}}\text{—(CH}_2)_4\text{—}\overset{\text{O}}{\overset{\|}{\text{C}}}\text{—NH(CH}_2)_6\text{—NH—}\right]_n + 2\,n\text{H}_2\text{O}$$

Nylon 6, 6

2. Polyester

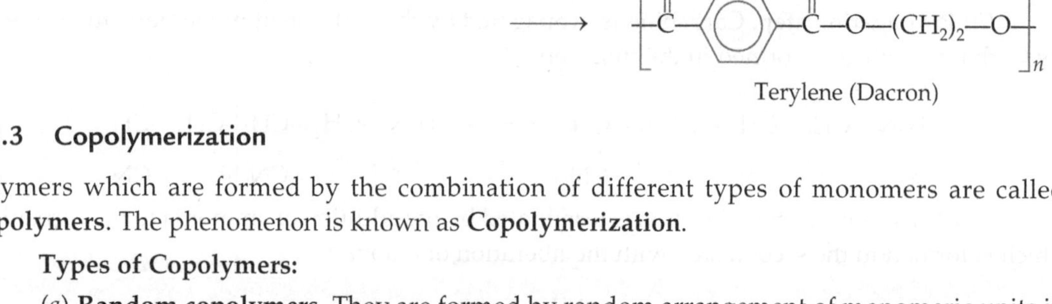

Terephthalic acid — Ethylene glycol

Terylene (Dacron)

4.2.3 Copolymerization

Polymers which are formed by the combination of different types of monomers are called **Copolymers**. The phenomenon is known as **Copolymerization**.

Types of Copolymers:

(*a*) **Random copolymers.** They are formed by random arrangement of monomeric units in the chain.

—A—B—A—A—B—A—B—B—B—A—A—

(*b*) **Alternate copolymers.** They are formed by alternative arrangement of monomeric units in the chain.

—A—B—A—B—A—B—A—B—A—

(*c*) **Block copolymer.** It consists of blocks of repeating units of one type alternating with blocks of another type.

—A—A—A—A—B—B—B—B—A—A—A—A—

(*d*) **Graft copolymer.** It consists of a linear polymer chain of one type to which the side chain of a different type has been grafted.

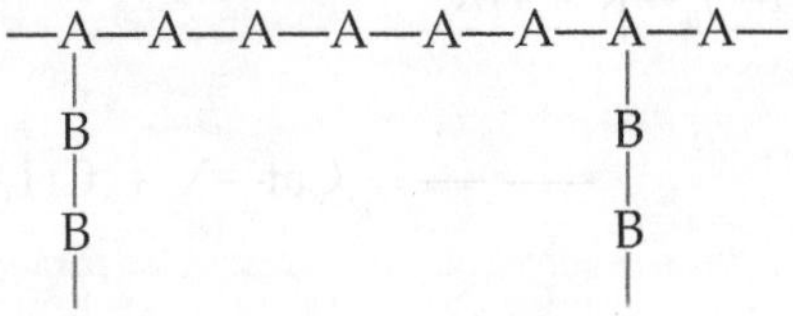

Example:

1. BUNA-S

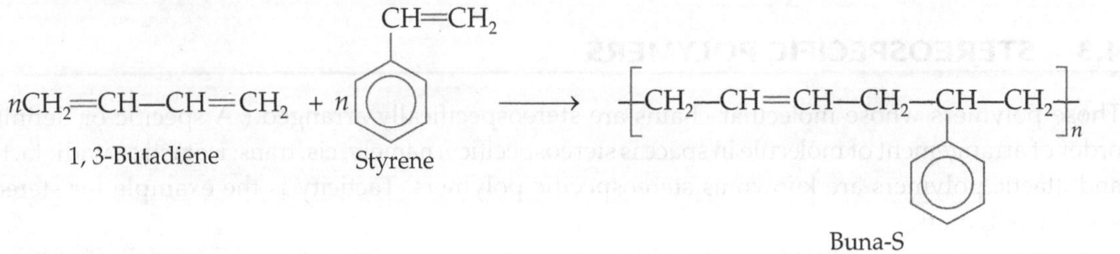

2. BUNA-N

$$nCH_2{=}CH{-}CH{=}CH_2 + nCH_2{=}\underset{}{CH(CN)} \xrightarrow{\text{Co-polymerization}} \left[CH_2{-}CH{=}CH{-}CH_2{-}CH_2{-}CH(CN) \right]_n$$

1,3-Butadiene Acrylonitrile Buna-N

4.2.4 Ziegler-Natta Catalysis or Co-ordination Polymerization

Ziegler (1953) and Natta (1955) discovered that in the presence of a combination of a transition metal halide (like $TiCl_4$ or $TiCl_3$, $ZnBr_3$, $TiCl_2$, halides of V, Zr, Cr, Mo and W) with an organometallic compounds (like triethylaluminium, trimethyl aluminum), Stereo specific polymerization is carried out. In this polymerization Ziegler-Natta catalyst is used so, this is also known as **Ziegler-Natta catalysis** or also called as co-ordination polymerization.

Mechanism:

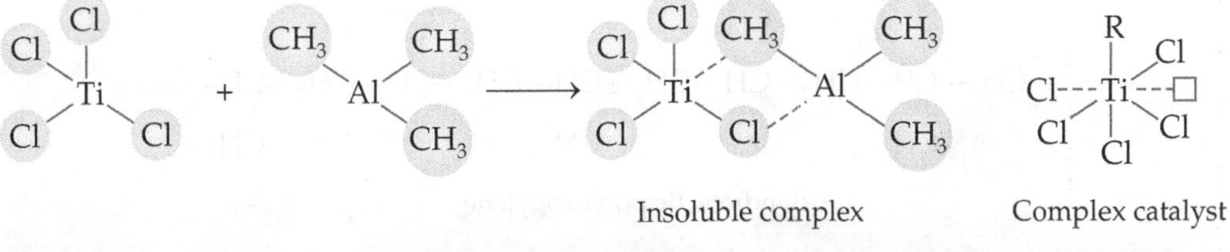

Initiation:

$$\underset{\text{(Complex catalyst)}}{Cat{-}R'} + \underset{\text{Monomer}}{CH_2{=}\underset{|\atop R}{CH}} \longrightarrow Cat{-}CH_2{-}\underset{|\atop R}{CH}{-}R'$$

Propagation:

$$Cat{-}CH_2{-}\underset{|\atop R}{CH}{-}R' + nCH_2{=}\underset{|\atop R}{CH} \longrightarrow Cat{-}CH_2{-}\underset{|\atop R}{CH}{-}({-}CH_2{-}\underset{|\atop R}{CH}{-})_n{-}R'$$

Termination (with active hydrogen compound):

$$Cat{-}CH_2{-}\underset{|\atop R}{CH}{-}({-}CH_2{-}\underset{|\atop R}{CH}{-})_n{-}R + HX$$

$$\longrightarrow Cat{-}X + CH_3{-}\underset{|\atop R}{CH}{-}({-}CH_2{-}\underset{|\atop R}{CH}{-})_n{-}R$$

Ziegler-Natta catalysis or polymerization is used for the preparation of polypropylene, polyethylene, polydiene etc.

4.3 STEREOSPECIFIC POLYMERS

Those polymers whose molecular chains are stereospecifically arranged (A specific or definite order of arrangement of molecule in space is stereospecific), namely, cis, trans, isotactic, syndiotactic and atactic polymers are known as stereospecific polymers. Tacticity is the example for stereo-specific polymers.

Tacticity. The orientation of monomeric units in a polymer molecule can take place in an orderly or disorderly fashion with respect to the main chain. The difference in configuration (tacticity) do affect their physical properties. Depending on the tacticity, there are three different types of polymers.

1. Isotactic polymers. The polymers in which the functional groups are all on the same side of the chain are known as Isotactic polymers.

Example. Polypropylene

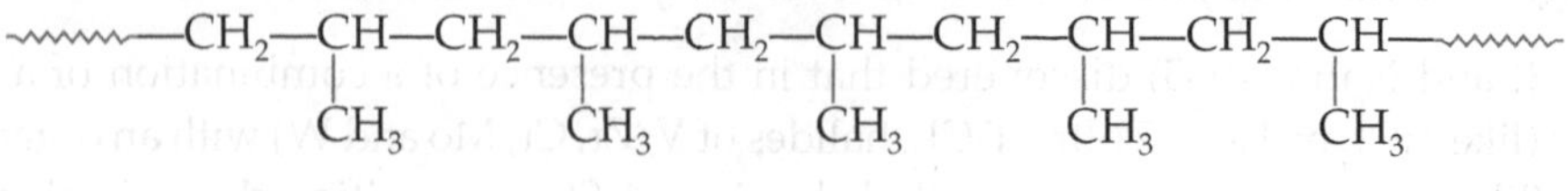

Isotactic polypropylene

2. Syndiotactic polymers. The polymers in which the functional groups are arranged alternatively are known as syndiotactic polymers.

Example. Polypropylene

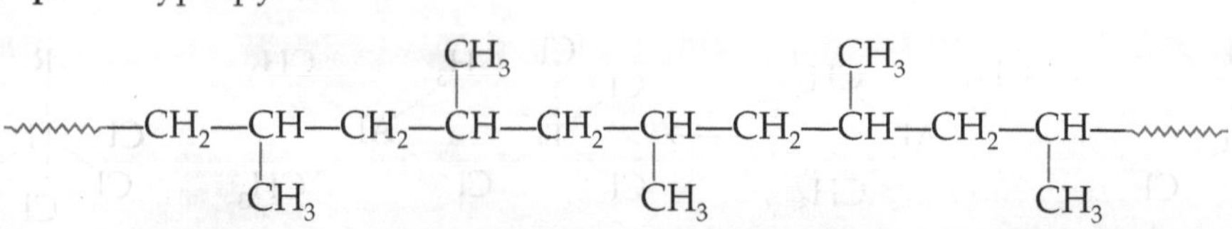

Syndiotactic polypropylene

3. Atactic polymers. The polymers in which the arrangements of functional groups are at random around the main chain are known as atactic polymers.

Example. Polypropylene

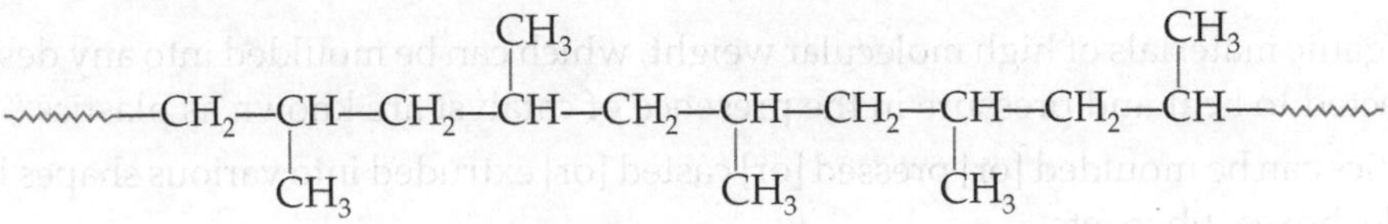

Atactic polypropylene

4.4 PHYSICAL AND MECHANICAL PROPERTIES OF POLYMERS

Physical properties. The physical properties of a polymer, such as its strength and flexibility depend on:

- **Chain length.** In general, the longer the chains the stronger the polymer;
- **Side groups.** Polar side groups give stronger attraction between polymer chains, making the polymer stronger;
- **Branching.** Straight, unbranched chains can pack together more closely than highly branched chains, giving polymers that are more crystalline and therefore stronger;
- **Cross-linking.** If polymer chains are linked together extensively by covalent bonds, the polymer is harder and more difficult to melt.

For example. Polystyrene, Polyvinyl acetate and polymethyl methacrylate are typically amorphous.

Mechanical properties. The bulk properties of a polymer are those most often of end-use interest. These are the properties that dictate how the polymer actually behaves on a macroscopic scale.

Tensile strength. The tensile strength of a material quantifies how much stress the material will endure before suffering permanent deformation. This is very important in applications that rely upon a polymer's physical strength or durability. For example, a rubber band with a higher tensile strength will hold a greater weight before snapping. In general, tensile strength increases with polymer chain length and crosslinking of polymer chains.

Young's modulus of elasticity. Young's Modulus quantifies the elasticity of the polymer. It is defined, for small strains, as the ratio of rate of change of stress to strain. Like tensile strength, this is highly relevant in polymer applications involving the physical properties of polymers, such as rubber bands. The modulus is strongly dependent on temperature. Visco-elasticity describes a complex time-dependent elastic response, which will exhibit hysteresis in the stress-strain curve when the load is removed. Dynamic mechanical analysis or DMA measures this complex modulus by oscillating the load and measuring the resulting strain as a function of time.

4.5 PLASTICS

The Word Plastic is derived from Greek Word 'Plastikos' meaning capable of being moulded or shaped.

"Organic materials of high molecular weight, which can be moulded into any desired form, when subjected to heat and pressure in the presence of catalyst are known as plastics."

Plastics can be moulded [or] pressed [or] casted [or] extruded into various shapes like plates, tubes, bottle, boxes, fibers etc.

Plastic materials have

- Sufficient rigidity
- Mechanical strength
- Insulation property
- Low density
- Low cost
- Light weight
- Resistance to chemicals, corrosion, insect abrasion etc. to shape them into desired article.

4.6 TYPES OF PLASTICS

On the basis of their physical characteristics, plastics are usually divided into three types:

1. Thermoplastics.
2. Thermosets.
3. Elastomers.

1. Thermoplastics. Thermoplastics have a linear or branched molecular structure which determines their strength and thermal behaviour; they are flexible at ordinary temperature. At approx 120 – 180°C, thermoplastics become a pasty/liquid mass. The service temperature range for thermoplastics is considerable lower than that for thermosets.

- These are prepared in single step process.
- These plastic are generally amorphous or crystalline.
- These are linear or branched polymers with weak intermolecular attractive forces.
- These plastics become soft on heating and stiffen on cooling at any time.

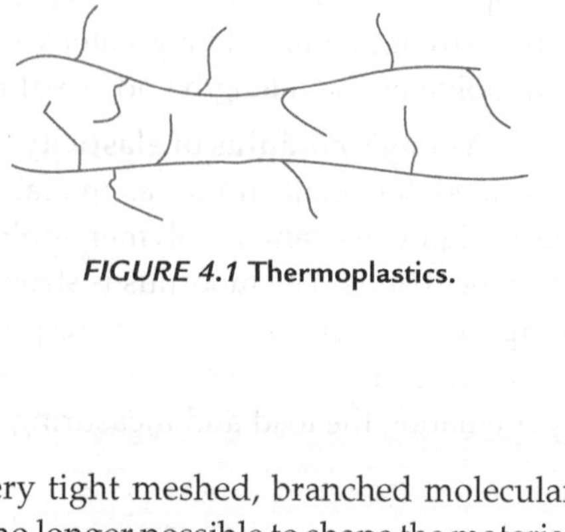

FIGURE 4.1 **Thermoplastics.**

- They may be remoulded, reshaped and reused.
- They can be recycled.

Example. PE, PS, PVC, Nylons, Teflon etc.

2. Thermosets. Thermosets are hard and have a very tight meshed, branched molecular structure. Curing proceeds during shaping, after which it is no longer possible to shape the material by heating. Further shaping may then only be performed by machining.

- They are normally produced.
- They are cross-linked, three dimensional structures.
- Once a product is prepared by heating and moulding, it can't be changed by further heating.
- On excess heating materials undergoes destruction but not become soft.

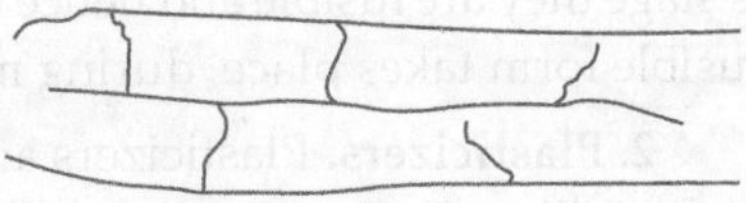

FIGURE 4.2 **Thermosettings.**

Example. Bakelite.

3. Elastomers. Elastomers also have a cross-linked structure; with looser mesh than thermosets, giving rise to a degree of elasticity. Once shaped, elastomers also cannot be reshaped by heating. Elastomers are used to produce automobile tyres.

Differences between thermosettings and thermoplastics

Thermoplastics	Thermosettings
1. Formed by either addition or condensation polymerization.	Formed by condensation polymerization.
2. They are linear or branched linked structures.	They are cross-linked or three dimensional structures.
3. Polymer chains are held together by weak attractive forces.	Polymer chains are held together by covalent cross-links.
4. They soften on heating and stiffen on cooling.	They don't soften on heating.
5. They are soluble.	They are insoluble.
6. They can be remoulded, reshaped and reused.	They cannot be remoulded.
7. They can be recycled.	They cannot be recycled.
8. They are tough.	They are brittle.
9. They have low melting points.	They have high melting points.
10. They have usually linear structures. Example, Polyethylene, PVC, Teflon.	They have three dimensional cross-linked structures. Example, Bakelite, Urea-formaldehyde resin.

4.7 COMPOUNDING OF PLASTICS OR MOULDING CONSTITUENTS OF A PLASTIC

Usually the high polymer is mixed with 4 to 10 ingredients, each of which either discharges a useful property to the finished artifact. This is called mix. The main types of compounding ingredients and their functions are described below.

1. Resins. Resin is the binder, which holds the different constituents together. Thermosetting resins are, usually, supplied as linear polymers of comparatively low molecular weight, because at this stage they are fusible and hence mouldable. The conversion of this fusible form into cross-linked infusible form takes place, during moulding itself, in presence of catalyst etc.

2. Plasticizers. Plasticizers are materials that are added to resins to increase their plasticity and flexibility. Their action is considered to be the result of the neutralization of part of the inter molecular forces of attraction between macromolecules of resins. Thus they impact greater freedom of movement between the polymeric macromolecules of resin, thereby increasing the flexibility and plasticity of the compound material, but at the same time reducing its strength and also decreasing its chemical resistance.

Example. Vegetable oils, camphor, esters (of steric or oleic acids) and some phosphates (tricresyl phosphates, tributyl phosphates) etc.

3. Fillers. These are added to give to the final plastic better hardness, tensile strength, opacity, finish and workability characters to the products.

Example:

(*i*) Carborandum, quartz and mica are added to provide extra hardness;

(*ii*) Barium salts as are added to make the plastic impervious to X-rays;

(*iii*) Addition of asbestos provides heat and corrosion-resistances.

The percentage of filler varies and can be up to 50% of the total moulding mixture.

4. Lubricants. Lubricants like Waxes, oils, stearates, oleates and soaps are employed to make the moulding of plastic easier and to impart a flawess, glossy finish to the products. Lubricants prevent the plastic material from sticking to the fabricating equipment.

5. Catalysts or accelerators. Catalysts or accelerators are added only in the case of thermosetting plastics with the object of accelerating the polymerization of fusible resin, during moulding operation, into cross-linked infusible form.

Example. Catalysts used are hydrogen peroxide, benzoyl peroxide, acetylsulphuric acid, metals like Ag, Cu and Pb, Metallic oxides such as ZnO, ammonia and its salts.

6. Stabilizers. The purpose of adding stabilizers is to improve the thermal stability during processing. For example, vinylchloride and vinylidine chloridepolymers show a tendency to undergo decomposition and discoloration at moulding temperatures. So during their moulding heat stabilizers are used.

Example:

(*a*) Opaque moulding compounds like salts of lead, *e.g*, white lead, lead chromate, litharge, red lead, lead silicate and lead naphthenate.

(*b*) Transparent moulding compounds like stearates of lead, Cadmium and Barium.

7. Colouring Materials (Antioxidants). Colour appeal is often of prime importance in high polymer artifacts. The main colouring materials are

Example. Organic dyestuffs and opaque inorganic pigments.

4.8 MOULDING OF PLASTICS INTO ARTICLES OR FABRICATION OF PLASTICS

Many methods of fabricating plastics into desired shaped articles are employed. The methods, usually, depend upon the type of resins used, *i.e.*, whether thermosetting or thermoplastic. Commonly used plastic fabrication methods are described below.

4.8.1 Compression Moulding

Compression moulding is a method applied to both thermoplastic and thermosetting resins. The predetermined quantity of plastic ingradients in proper proportions get filled between the two half pieces of mould. Which are capable of being moved relative to each other? Heat and pressure are then applied according to specifications. The cavities get filled with fluidized plastic. Two halves are closed very slowly. Finally, curing is done either by heating (in case of thermosetting) or cooling (in case of thermoplastics). After curing the moulded article is taken out by opening the mould parts.

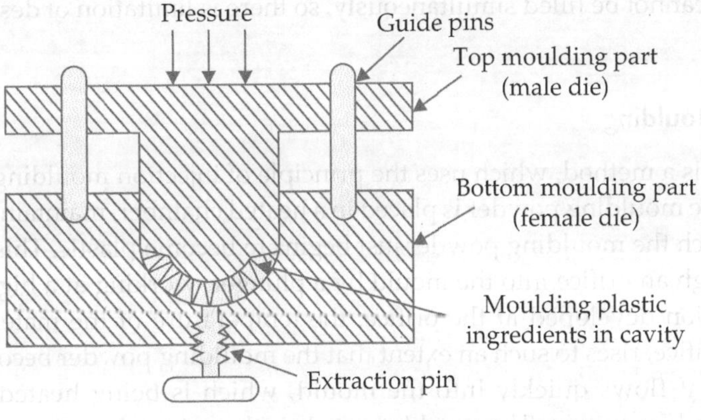

FIGURE 4.3 **Compression moulding of plastics.**

4.8.2 Injection Moulding

Injection Moulding is applicable mainly to thermoplastic resins. In this, the moulding plastic powder is fed into a heated cylinder, from where it is injected at a controlled rate into the tightly locked mould by means of a screw arrangement or by a piston plunger. The mould is kept cold to allow the hot plastic to cure and become rigid. When the materials have been cured sufficiently, half of the mould is opened to allow the injection of the finished article without any deformation, etc. Heating is done by oil or electricity.

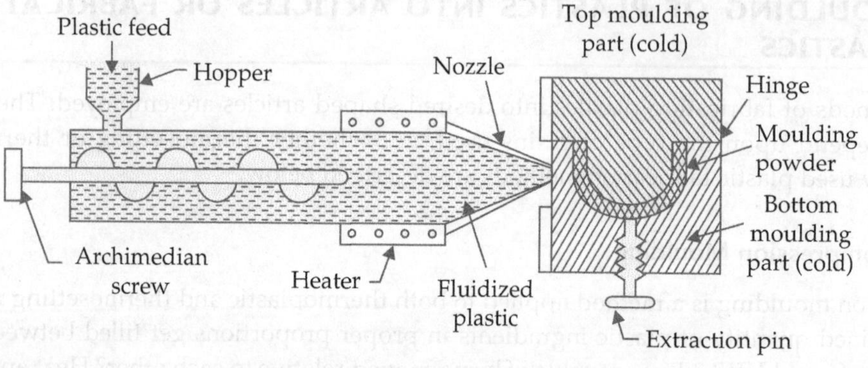

FIGURE 4.4 **Injection moulding of plastics.**

Advantages:

This method is the most widely used for moulding of thermoplastics because of high speed production, low mould cost, very low loss of material and low finishing cost. Because a large number of cavities cannot be filled simultaneously, so there is limitation of design of articles to be moulded.

4.8.3 Transfer Moulding

Transfer moulding is a method, which uses the principle of injection moulding for thermosetting materials. In this, the moulding powder is placed in a heated chamber, maintained at the minimum temperature at which the moulding powder just begins to become plastic. This plastic material is then injected through an orifice into the mould by a plunger, working at a high pressure. Due to the very great friction developed at the orifice, the temperature of the material, at the time of ejection from the orifice, rises to such an extent that the moulding powder becomes almost liquid, and consequently, it flows quickly into the mould, which is being heated up to the curing temperature required for setting. The moulded article is then ejected mechanically.

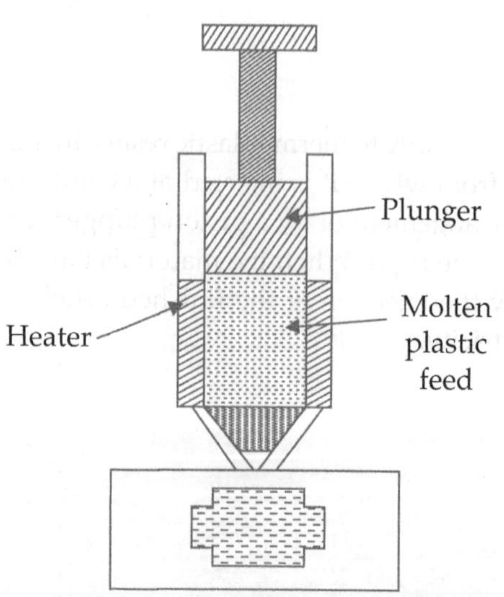

FIGURE 4.5 **Transfer moulding.**

Advantages:

1. In tricate shapes, not attainable by compression moulding can readily by produced.
2. Article produced is free from marks.
3. Finishing cost of fabricated article is almost entirely eliminated.
4. Mould cost is less since it involves very low abrasive action.

4.8.4 Extrusion Moulding

Extrusion moulding is used mainly for continuous moulding of thermoplastic materials into articles of uniform cross-section like tubes, rods, strips, insulated electric cables. The thermoplastic ingredients are heated to plastic condition and then pushed by means of a screw conveyor into a die, having the required outer shape. Of the article to the manufactured. Here the plastic mass gets cooled, due to the atmospheric exposure (or artificially by air jets). A long conveyor carries away continuously the cooled product.

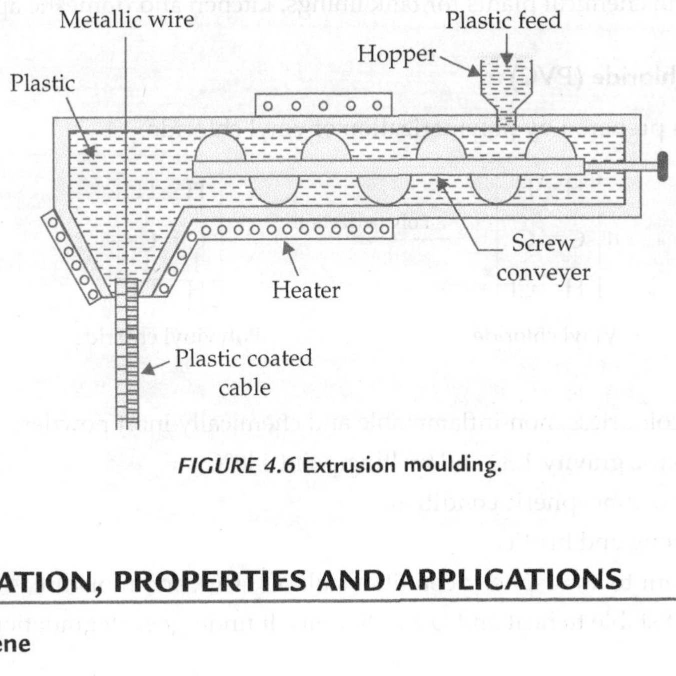

FIGURE 4.6 **Extrusion moulding.**

4.9 PREPARATION, PROPERTIES AND APPLICATIONS

4.9.1 Polyethylene

It is obtained by the polymerization of ethylene

$$nCH_2{=}CH_2 \longrightarrow -\!\!\!+ CH_2{-}CH_2 +\!\!\!-_n$$

Ethylene Polyethylene

Polyethylene is produced by the high pressure process has a branched structure and, therefore, is flexible and tough. On the other hand low pressure process result in a completely linear polyethylene, having their density and better chemical resistance.

The commercial polyethylene can be subdivided into three types:

1. Low density polyethylene (LDPE).
2. Medium density polyethylene (MDPE).
3. High density polyethylene (HDPE).

The high-density polyethylene, which is completely linear, has better chemical resistance, particularly to organic solvents, a higher softening point and greater rigidity than the medium or low density polyethylene, but it has low impact strength and is relatively brittle.

Properties:

1. Polyethylene is a rigid and waxy.
2. It is translucent, non-polar material.
3. It exhibits chemical resistance to strong acids, alkalis and salt solutions at room temperature.
4. It is good insulator of electricity.
5. It is swollen and permeable to most oils and organic solvents, particularly kerosene.

Uses:

1. Poly ethylene used for making insulating parts.
2. It is used for making bottle caps, packing materials, tubes, coated wires.
3. It is used in chemical plants for tank linings, kitchen and domestic appliances.

4.9.2 Polyvinyl Chloride (PVC)

Preparation. PVC is prepared by polymerization of vinyl chloride.

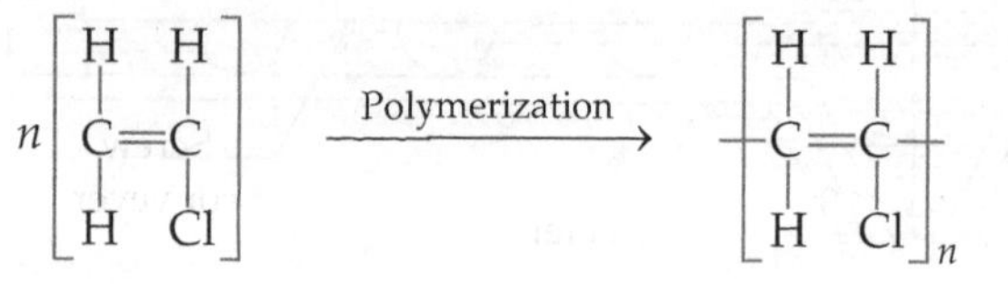

Vinyl chloride   Polyvinyl chloride

Properties:

1. PVC is a colourless, non-inflammable and chemically inert powder.
2. It has specific gravity 1.33 and melting point 148°C.
3. Resistant to atmospheric condition.
4. PVC is strong and brittle.
5. It is resistant to atmospheric conditions like O_2, CO_2 and moisture.
6. PVC is not stable to heat and UV radiations. It undergoes degradation.

Applications:

1. PVC treated with plasticizers is called plasticized.
2. PVC used for electrical insulations, injection moulding articles like tool handles.
3. It is used radio, telephone components, chemical containers, and foams used in buildings.
4. Unplasticized PVC is rigid, highly resistant to chemicals and brittle.
5. PVC is used for making safety helmets, refrigerator components, tyres, cycle and motor cycle mudguards.

4.9.3 Bakelite (Phenol-Formaldehyde Resin)

Preparation:

Bakelite is prepared by condensing phenol with formaldehyde in presence of acidic/alkaline catalyst.

Step I. Phenol reacts with formaldehyde produce o-Hydroxy methyl phenol and p-Hydroxy methyl phenol.

$$CH_2{=}O + H^{\oplus} \longrightarrow \overset{\oplus}{C}H_2{-}OH$$

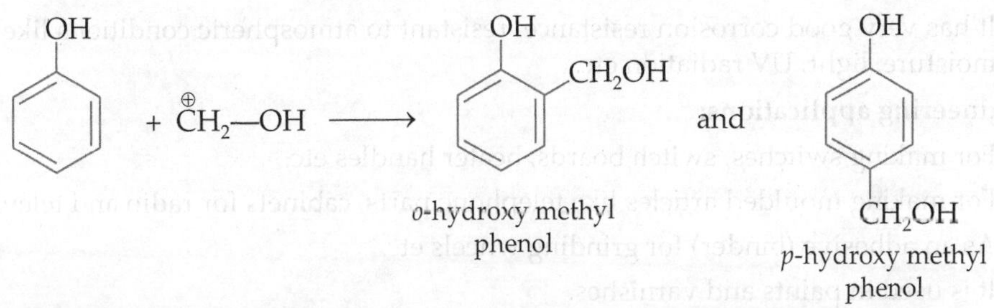

Step II. *o*-Hydroxy methyl phenol and phenol, which reacts to form linear polymer Novalac resin.

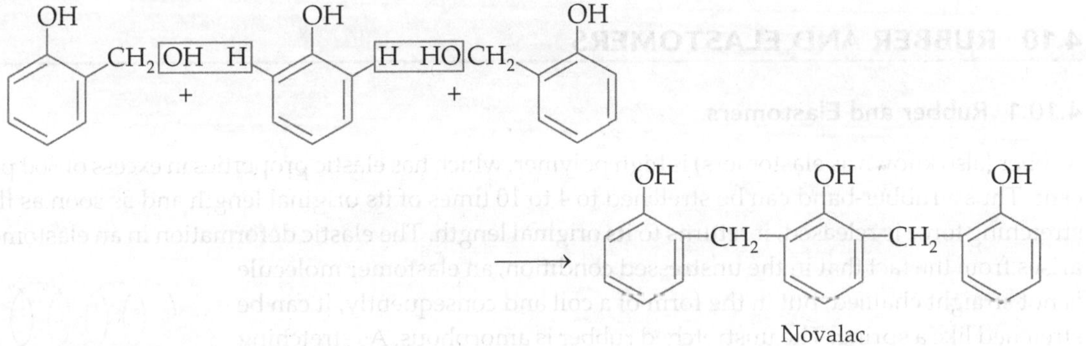

Step III. During moulding, catalyst hexamethylene tetraamine $[(CH_2)_6N_4]$ is added. Addition of this catalyst provides excess of formaldehyde, which converts the soluble and fusible Novalac resin into hard, infusible and insoluble solid of cross-linked polymer Bakelite.

$$[(CH_2)_6N_4] + H_2O \longrightarrow 6CH_2O + 4NH_3$$

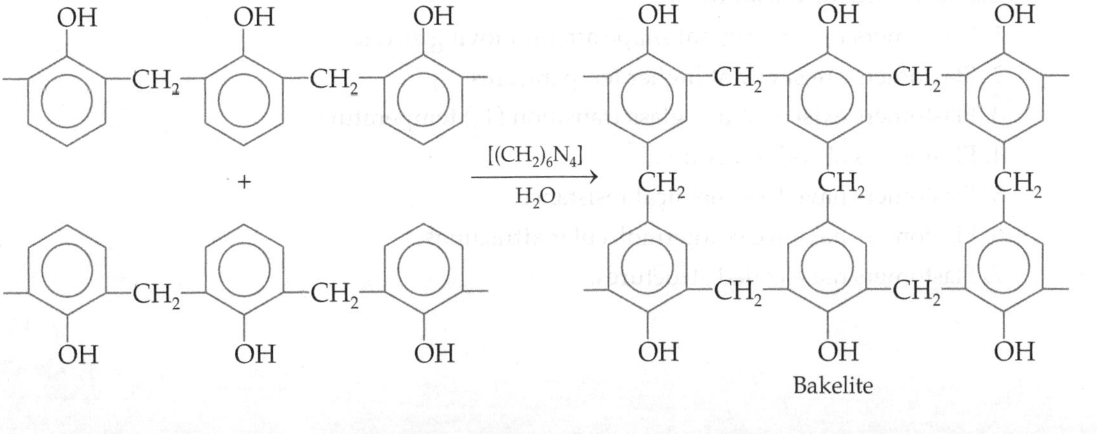

Properties:

1. It is hard, rigid and strong.
2. It is a scratch resistant and water resistant polymer.
3. Bakelite has got good chemical resistance, resistant to acids, salts and many organic solvents, but it is attacked by alkalis due to the presence of —OH groups.
4. It is a good anion exchanging resin, exchanges —OH group with any other anion.
5. It is an excellent electrical insulator.
6. It is a very good adhesive.
7. It has very good corrosion resistance, resistant to atmospheric conditions like O_2, CO_2, moisture, light, UV radiation etc.

Engineering applications:

1. For making switches, switch boards, heater handles etc.
2. For making moulded articles like telephone parts, cabinets for radio and television.
3. As an adhesive (binder) for grinding wheels etc.
4. It is used in paints and varnishes.
5. For making bearings used in propeller shafts, paper industry and rolling mills.

4.10 RUBBER AND ELASTOMERS

4.10.1 Rubber and Elastomers

Rubber (also known as elastomers) is high polymer, which has elastic properties in excess of 300 per cent. Thus a rubber-band can be stretched to 4 to 10 times of its original length and as soon as the stretching force is released, it returns to its original length. The elastic deformation in an elastomer arises from the fact that in the unstressed condition, an elastomer molecule is not straight chained, but in the form of a coil and consequently, it can be stretched like a spring. The unstretched rubber is amorphous. As stretching is done, the macro molecules get partially aligned with respect to another, thereby causing crystallization and consequent, stiffening of material. On releasing the deforming stress, the chains get reverted back to their original coiled state and the material again becomes amorphous.

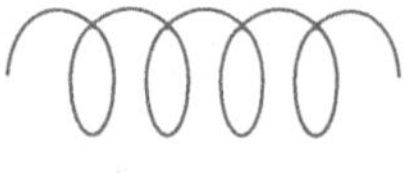

FIGURE 4.7

Characteristics of Elastomers:

1. Elastomers regain original shape after removing stress.
2. Elastomers are sticky at higher temperatures.
3. Elastomers exist at above glass transition (T_g) temperature.
4. Elastomers absorb moisture.
5. Elastomers have less chemical resistance.
6. Elastomers have weak intermolecular attractions.
7. Elastomers have coiled structures.

Preparation of Natural Rubber. Natural rubber consists of basic material latex, which is a dispersion of isoprene. During the treatment, these isoprene molecules polymerize to form, long coiled chains of cis-poly isoprene.

Isoprene — Poly isoprene

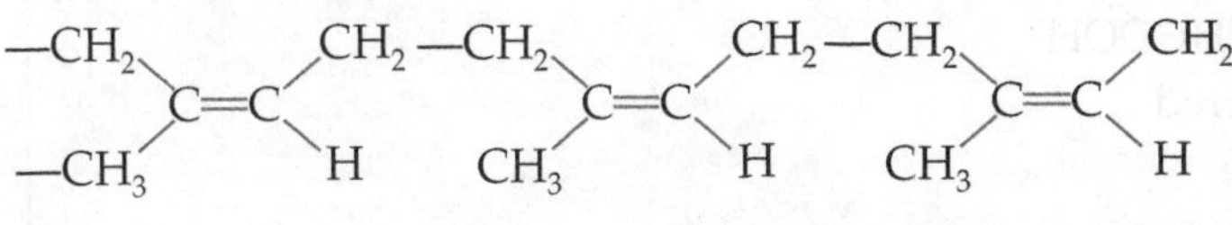

Structure of natural rubber (cis-poly isoprene)

The molecule weight of raw rubber is about 100,000–150,000. Natural rubber is made from the saps of plants like have a brasillians and guayule, found in tropical and semi-tropical countries. The rubber latex (or milky-liquid rubber) is obtained by making incisions in the bark of the rubber trees and allowing the saps to flow out into small vessels. Tapping is usually, done at intervals of about six months. The latex is emptied into buckets and transferred to a factory for treatment.

Processing of Latex. Latex is diluted to contain between 15 to 20% of rubber and filtered to eliminate any dirt present in it. Acetic acid or formic acid is then added, when rubber is coagulated to soft white mass. The coagulated mass is washed. The coagulum is treated as follows.

(*a*) **Crepe rubber.** The coagulum is allowed to drain for about 2 hours. Then, passed through a creping machine which consists essentially of two rollers with longitudinal grooves, upon which water is sprayed. The spongy coagulated in to a sheet, which possesses an uneven rough surface resembling crepe paper.

(*b*) **Smoked rubber.** Coagulation is carried out in long tanks having sides with vertical grooves, fitted with metal plates. Diluted latex is poured into the tanks, with plates removed. Then, dilute acetic or formic acid is added and the mixture is stirred thoroughly. The partition plates are then inserted into the grooves. The tanks are kept undisturbed for about 16 hours. The tough slabs of coagulum so formed between the partition plates are removed. The slabs are then passed through a series of smooth rollers., with decreasing clearance in between them and water is sprayed simultaneously at the center of the rollers. The final rollers has a clearness of such a design so as to give ribbed pattern to the final rubber sheet. Ribbed–pattern on the surface of rubber sheet exposes greater surface, and this facilitates consequent drying and also prevents the sheets from adhering together on stacking. The sheets are then hung for about 4 days in a smoke house, in which temperature between 40 to 50°C is maintained. The crude rubber thus–obtained is amber in colour and translucent.

Flow Chart of Processing of Latex:

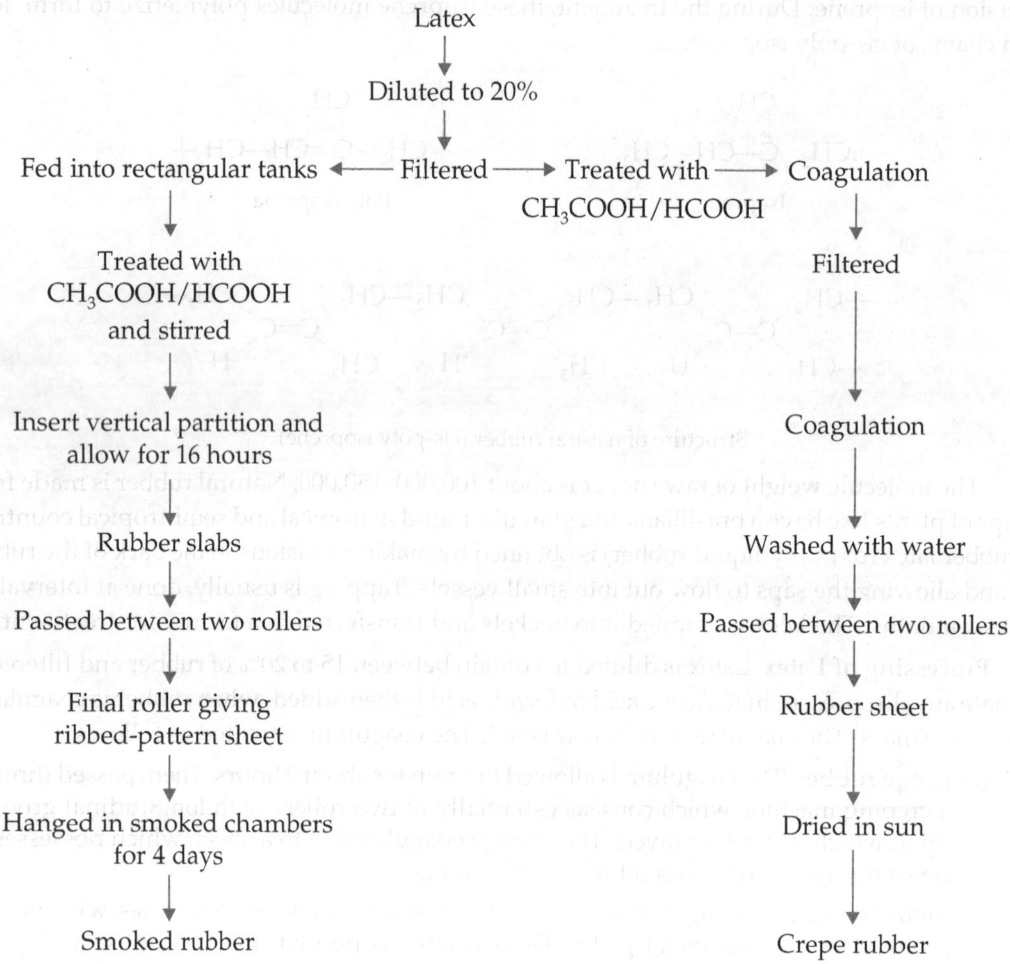

Gutta-percha. Gutta-percha is obtained from the mature leaves of dichopsis gutta and palagum gutta trees (belonging to sapetaceae family), grown mostly in Broneo, Malaya, Sumatra, Islands. Gutta-percha may be recovered by solvent extraction, when insoluble resins and gums are separated. Alternatively, the mature leaves are ground carefully, treated with water at about 70°C for half an hour and then poured into cold water, when gutta-percha floats on water surface and is removed.

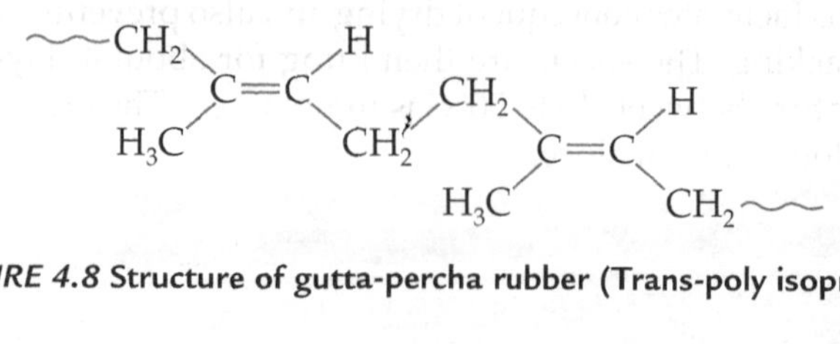

FIGURE 4.8 **Structure of gutta-percha rubber (Trans-poly isoprene).**

Properties:

1. At room temperature, gutta-percha is horny and tough, but it softens and becomes tacky at about 100°C.
2. It is soluble in aliphatic hydrocarbons, but insoluble in aromatic and chlorinated hydrocarbons. Structurally it is trans-poly isoprene, whereas havea rubber is cis-isomer.

Uses. In the manufacture of golf ball covers, submarine cables, adhesives and tissues for surgical purposes.

4.11 DRAW BACKS OF RAW RUBBER

Raw rubber has the following draw backs:

1. **It is plastic in nature.** It becomes soft at high temperature and is too brittle at low temperature. So it can be used in the temperature range of 10–60°C only.
2. **It is weak.** Its tensile strength is only 200 kg/cm^2.
3. It has large water absorption capacity.
4. It is non-resistant to non-polar solvents like vegetable and mineral oils, gasoline, benzene and CCl_4.
5. It is attacked by oxidizing agents like HNO_3, conc. H_2SO_4, NaOCl, ClO_2 and Cr_2O_3 etc.
6. It perishes, due to oxidation in air.
7. It swells considerably in organic solvents and gradually disintegrates.
8. It posses marked thickness, *i.e.*, when two fresh raw rubber surfaces are pressed together; they coalesce to form a single piece.
9. It has little durability.
10. When stretched to a great extent, it suffers permanent deformation, because of the sliding or slippage of some molecular chains over each other

4.12 VULCANIZATION (DISCOVERED BY CHARLES GOODYEAR IN 1839)

To improve the properties of rubber, it is compounded with some chemicals like sulphur, hydrogen sulphide, benzoyl chloride etc. Most important is addition of sulphur. The process consists in heating the raw rubber with sulphur to 100–140°C. The added sulphur combines chemically at the double bonds of different rubber springs. Vulcanization thus serves to stiffen the material by a sort of anchoring and consequently, preventing intermolecular movement of rubber springs. The extent of stiffness of vulcanized rubber depends on the amount of sulphur added. For example, a tyre rubber may contain 3–5% of sulphur, but a battery case rubber may contain as much as 30% sulphur.

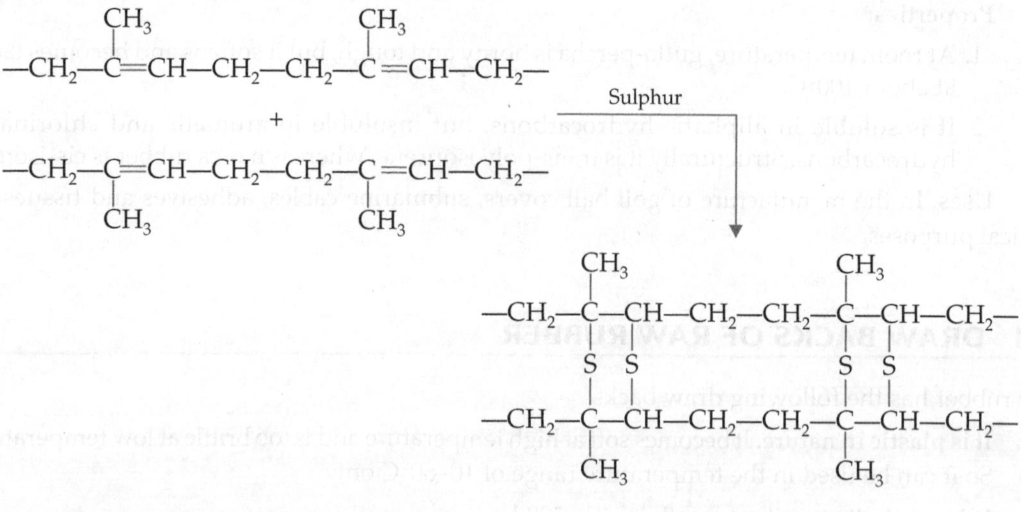

Vulcanized rubber

Vulcanite or Ebonite or Hard rubber. Vulcanite or Ebonite or Hard rubber contains about 32% sulphur. The cross-linking of the isoprene spirals is attained to such an extent that vulcanite has practically no elasticity. Thus, vulcanite is tough, high abrasion-resistant, can be moulded, sawn, drilled and polished. It has extremely good electrical insulation properties.

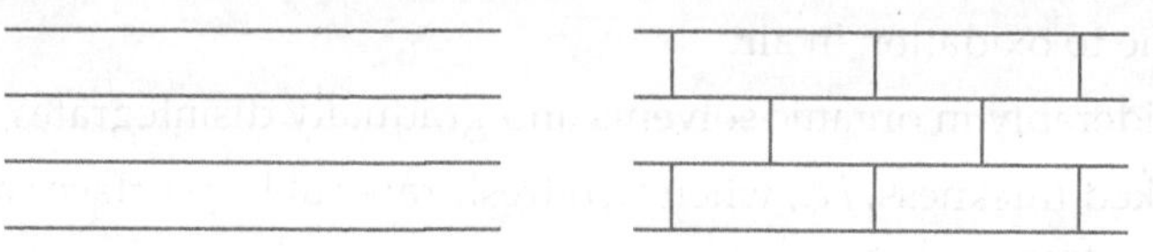

Before vulcanization   After vulcanization

FIGURE 4.9

Advantages of vulcanization (properties):

1. Vulcanized rubber has good tensile strength and local bearing capacity.
2. Vulcanized rubber has good elasticity depending on the extent of vulcanization.
3. It posses low water absorption tendency.
4. Vulcanized rubber has higher resistance to oxidation and atmospheric gases like O_2, CO_2, moisture, light and U.V radiation.
5. Vulcanized rubber is a good electrical insulator and ebonite is a best electrical insulator.
6. The abrasion resistance of natural rubber increases tremendously.
7. The resistance towards organic solvents and chemicals increase.
8. Vulcanized rubber posses good impact resistance.
9. Vulcanized rubber posses very good resilience *i.e.*, the article made from it returns to original shape, when the deforming load is removed.

4.13 METHODS FOR IMPROVING QUALITY OF RUBBER

(*a*) **Mastication:**

1. Making rubber into small pieces by mechanical stirring is known as mastication.
2. Rubber is converted into soft and semi solid mass.
3. CH_2—CH_2 links in the rubber are weak and they undergo cleavage on mechanical stirring.
4. After mastication various chemicals are added to improve the qualities.

(*b*) **Compounding of rubber.** To improve the qualities from rubber, it is mixed with various chemicals like plasticizers, stabilizers, fillers, vulcanizing agents etc. is called compounding of rubber.

1. Vulcanization of rubber:

- To improve the undesirable properties of rubber, it is heated in the presence of sulphur.
- Heating of rubber in presence of sulphur is called vulcanization.
- On heating three dimensional cross-linkings of the chains through sulphur atoms formed.
- Rubber becomes stiff.
- This process converts rubber into insoluble, infusible thermosetting polymer.

2. Accelerators:

- Vulcanization process is catalyzed by adding accelerators.
- They reduce the time required for vulcanization.
- Accelerators may be inorganic or organic.
  - ➢ **Inorganic.** Lime, Magnesia, ZnO
  - ➢ **Organic.** Aldehydes, Amines, thio carbonates
- 0.5 to 1% of the accelerator is used.

3. Antioxidants:

- Antioxidants protect the rubber from oxidation by O_2, ozone and light.
  - ➢ **Example.** Phenylnaphthylamine, Phenolic substances, Phosphate.

4. Reinforcing agents:

- They provide strength, rigidity and toughness to the rubber.
  - ➢ **Example.** Carbon black, ZnO, $MgCO_3$, $CaCO_3$ and clays.

5. Fillers:

- The main function of the fillers is to change the physical properties and to reduce the cost.

6. Plasticizers:

- These are added to impart greater tenacity and adhesion to the rubber.
  - ➢ **Example.** Vegetable oils, Waxes, Stearic acid, Rosin etc.

7. Colouring agents:

- These are added to impart the desired colour to rubber.
  - ➢ TiO_2, ZnS, $BaSO_4$ – White
  - ➢ $PbCrO_4$ – Yellow

- ➢ Ferric oxide – Red
- ➢ Sb_2S_3 – Crimson red
- ➢ Ultramarine – Blue

8. **Miscellaneous agents:**
   - Baking soda is added to prepare sponge rubber.
   - Silica and pumice are added to cause abrasiveness.
   - Processed rubber is converted into various articles by calendaring or injection moulding or compressed moulding.

4.14 PREPARATION, PROPERTIES AND APPLICATIONS

4.14.1 SBR (Buna-S or Styrene-Butadiene Rubber)

Buna-S is also called as Styrene rubber or GR-S (Government Rubber Styrene) or Ameripol. Buna-S stands for the composition of the monomers and catalyst.

Bu stands for Butadiene-Monomer

Na stands for Sodium-Catalyst

S stands for Styrene-Monomer

Buna-S is produced by the copolymerization of butadiene with styrene using sodium as catalyst.

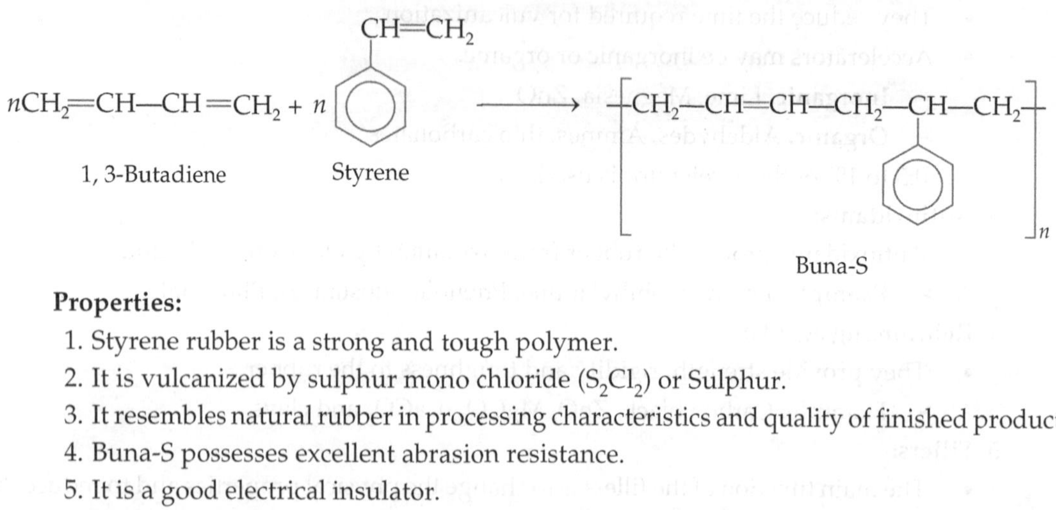

Properties:

1. Styrene rubber is a strong and tough polymer.
2. It is vulcanized by sulphur mono chloride (S_2Cl_2) or Sulphur.
3. It resembles natural rubber in processing characteristics and quality of finished products.
4. Buna-S possesses excellent abrasion resistance.
5. It is a good electrical insulator.
6. It is resistant to chemicals but swells in oils and attacked by even traces of ozone present in the atmosphere.

Engineering applications (Uses):

1. Major application of styrene rubber is in the manufacture of tyres.
2. It is used in the footwear industry for making shoe soles and footwear components.
3. It is also used for making wires and cable insulations.
4. The other applications of styrene rubber are for the production of floor files, tank linings in chemical industries and as adhesives.

4.14.2 Polysulphide Rubber (Thiokol Rubber)

Polysulphide rubber is made by the reaction between sodium polysulphide and ethylene dichloride.

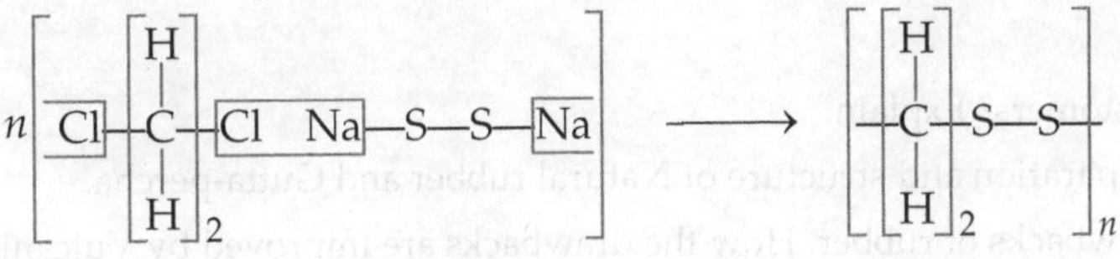

1, 2-Dichloroethane   Sodium polysulphide   Thiokol

Properties:

1. It is resistant to swelling.
2. It is resistant to oils.
3. It cannot be vulcanized and hence does not form a hard rubber.
4. It has poor heat resistance, poor abrasion resistance and low tensile strength.
5. It tends to lose its shape under continuous pressure.

Engineering applications (Uses):

1. Used to make sealants, gaskets, balloons, fabric coating.
2. Due to low permeability, it is used to make gasoline hoses.
3. The liquid polymers, when combined with oxidizers burn with great intensity and generate large volume of gas. Thus, it can be used in solid-fuel rocket propellants.

EXERCISE

1. Define the following terms:
   (*a*) Monomer
   (*b*) Polymer
   (*c*) Polymerization
   (*d*) Repeating unit
   (*e*) Degree of polymerization
   (*f*) Functionality.
2. Write about the following types of polymerizations with two examples for each:
   (*a*) Addition polymerization
   (*b*) Condensation polymerization
   (*c*) Co-polymerization
   (*d*) Ziegler-Natta catalysis or Co-ordination polymerization.
3. Write a note on Stereo specific polymers.
4. Explain the physical and mechanical properties of polymers.
5. What are plastics? Explain about thermoplastics and thermosetting plastics.
6. Write the differences between thermoplastics and thermosettings.
7. Write a note on compounding of plastics.
8. Explain the following fabrication methods of plastics:
   (*a*) Compression moulding
   (*b*) Injection moulding
   (*c*) Transfer moulding
   (*d*) Extrusion moulding.

9. Write the preparation, properties and applications of the following:
   (*a*) Polyethylene
   (*b*) Polyvinyl Chloride
   (*c*) Bakelite.
10. What are elastomers? Explain.
11. Write the preparation and structure of Natural rubber and Gutta-percha.
12. Write the drawbacks of rubber. How the drawbacks are improved by Vulcanization?
13. What are the additives are added to improve the drawbacks of rubber other than Sulphur?
14. Write the preparation, properties and applications of the following:
   (*a*) Styrene butadiene rubber (*b*) Thiokol rubber.

Chapter 5 FUELS

5.1 INTRODUCTION

Fuel. "A combustible substance containing carbon as the main constituent, which on proper burning liberates large amount of heat, which can be used economically for domestic as well as industrial purposes is called *Fuel*".

5.2 CLASSIFICATION OF FUELS

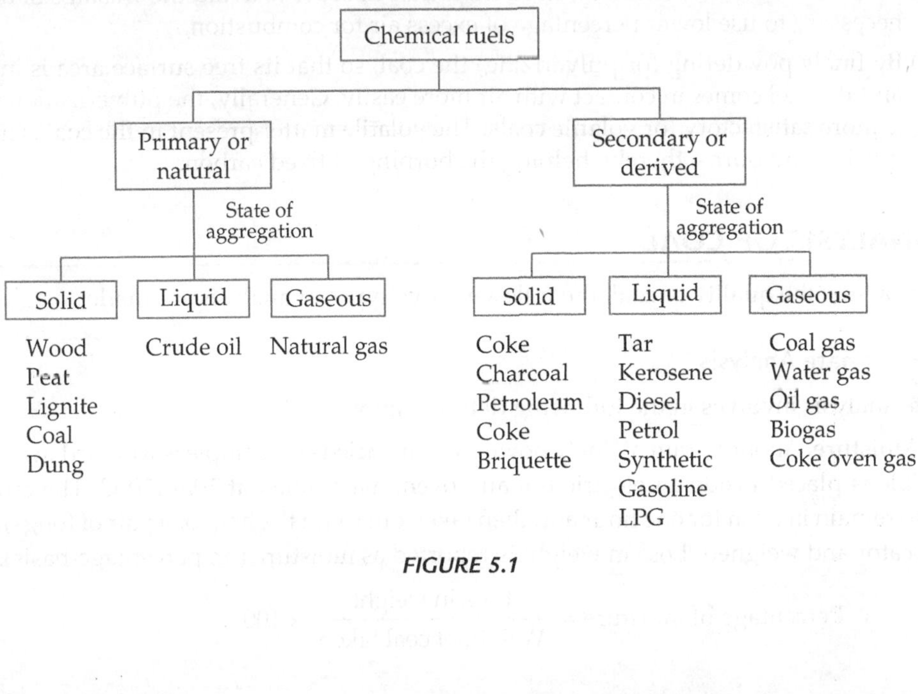

FIGURE 5.1

5.3 COMPARISON BETWEEN SOLID, LIQUID AND GASEOUS FUELS

Solid fuels	*Liquid fuels*	*Gaseous fuels*
Slow combustion and not easy to control it.	Quick combustion and can be controlled.	Combustion is rapid and burning can be controlled.
Transportation is difficult.	Transportation is easy through pipe lines.	Transportation is easy through pipe lines and containers.
Storage is safe.	There is risk in storing.	There is greatest risk in storing.
Calorific value is comparatively low.	Calorific value is comparatively higher.	Calorific value is highest.
Cannot be used in internal combustion engines.	Can be used in internal combustion engines.	Can be used in internal combustion engines to a lesser extent.
On burning ash and more smoke are produced.	No ash is produced but some flue gases are produced.	No ash no smoke are produced.

5.4 COAL

A highly carbonaceous matter that has been formed as a result of alteration of vegetable matter [*e.g.*, plants] under certain favourable conditions. It is chiefly composed of C, H, N and O besides non-combustible inorganic matter.

The combustion rate of solid fuels, like coal, is generally slow, because of the difficulty of through contact between the solid fuel and oxygen. The combustion rate can be increased by either of the following methods:

(*i*) By increasing the rate of oxygen supply, but this method involves wastage of large proportion of heat carried by air itself. Hence in order to avoid the wastage of heat, it is necessary to use lower percentage of excess air for combustion.

(*ii*) By finely powdering (or pulverizing) the coal, so that its free surface area is increased and the fuel comes in contact with air more easily. Generally, the pulverization method is more satisfactory for volatile coals. The volatile matter present in the coal is liberated quickly and burns, thereby helping the burning of fixed carbon.

5.5 ANALYSIS OF COAL

In order to assess the quality of coal, the following two types of analysis are made:

5.5.1 Proximate Analysis

Proximate analysis involves in the following determinations.

1. Moisture. About 1 gram of finely powdered air –dried coal sample is weighed in a crucible. The crucible is placed inside an electric hot air –oven, maintained at 105–110°C. The crucible is allowed to remain in oven for one hour and then taken out (with the help of a pair of tongs), cooled in a desiccator and weighed. Loss in weight is reported as moisture (on percentage-basis).

$$\text{Percentage of moisture} = \frac{\text{Loss in weight}}{\text{Weight of coal taken}} \times 100$$

2. Volatile matter. The dried sample of coal left in the crucible in (1) is then covered with a lid and placed in an electric furnace (muffle furnace), maintained at 925°C ± 20°C. The crucible is taken out of the oven after 7 minutes of heating. The crucible is cooled first in air, then inside a desiccator and weighed again. Loss in weight is reported as volatile matter on percentage-basis.

3. Ash. The residual coal in the crucible (2) is then heated without lid in a muffle furnace at 700°C + 50°C for half an hour. The crucible is then taken out, cooled first in air, then in desiccator and weighed. Heating, cooling and weighing is repeated, till a constant weight is obtained. The residue is reported as ash on percentage basis. Thus,

$$\text{Percentage of ash} = \frac{\text{Weight of ash left}}{\text{Weight of coal taken}} \times 100$$

4. Fixed carbon. Percentage of fixed carbon = 100 – Percentage of (moisture + volatile matter + ash)

Importance of proximate analysis. Proximate analysis provides following valuable information's in assessing the quality of coal.

***(i)* Moisture.** Moisture in coal evaporates during the burning of coal and it takes some of the liberated heat in the form of latent heat of evaporation. Therefore, moisture lowers the effective calorific value of coal. Moreover, it quenches the fire in the furnace. Hence, lesser the moisture content, better the quality of coal as a fuel. However, presence of moisture, up to 10% produces a more uniform "fuel-bed" and less of "fly-ash".

***(ii)* Volatile matter.** A high volatile matter content means that a high proportion of fuel will distill over as gas or vapour, a large proportion of which escapes unburnt, so higher volatile content in coal is undesirable. A high volatile matter containing coal burns with a long flame, high smoke and has low calorific value. Hence, lesser the volatile matter, better the rank of the coal. Moreover, the volatile matter affects the furnace volume and arrangement of heating space. Thus, high-volatile matter containing coals (usually of low rank) do not coke well; whereas medium-volatile matter content coals (containing 26–30% volatile matter) are capable of yielding hard and strong coke on carbonization. On the other hand, low-volatile matter containing coals do not coke at all and consequently, they are totally unsuitable for coke making.

***(iii)* Ash.** Ash is a useless, no combustible matter, which reduces the calorific value of coal. Moreover, ash causes the hindrance to the flow of air heat, thereby lowering the temperature. Also, it often causes trouble during firing by forming clinkers (*i.e.*, fused ash lumps), which block the interspaces of the great, on which coal is being burnt. This in-turn causes obstruction to air supply, thereby the burning of coal becomes irregular. Hence lower the ash content, better the quality of coal. The presence of ash (similar to moisture) also increases transporting, handling and storage costs. It also involves additional cost in ash disposal. The presence of ash also causes early wear of furnace walls, burning of apparatus and feeding mechanism.

***(iv)* Fixed carbon.** Higher the percentage of fixed carbon, greater is it's calorific and better the quality of coal. Greater the percentage of fixed carbon, smaller is the percentage of volatile matter. This also represents the quantity of carbon (in coal) that can be burnt by a primary current of air drawn through the hot bed of a fuel. Hence high percentage of fixed carbon is desirable. The percentage of fixed carbon helps in designing the furnace and the shape of the fire-box because it is the fixed carbon that burns in the solid state.

5.5.2 Ultimate Analysis

Ultimate analysis involves the following determinations.

1. Carbon and Hydrogen. About 1–2 gram of accurately weighed coal sample is burnt in a current of oxygen in a combustion apparatus. 'C' and 'H' of the coal are converted into CO_2 and H_2O respectively. The gaseous products of combustion are adsorbed respectively in KOH and $CaCl_2$ tubes of known weights. The increase in weights are then determined.

$$\underset{12}{C} + \underset{32}{O_2} \longrightarrow \underset{44}{CO_2}, \quad \underset{2}{H_2} + \underset{16}{\tfrac{1}{2}O_2} \longrightarrow \underset{18}{H_2O}$$

$$2KOH + CO_2 \longrightarrow K_2CO + H_2O + O_2$$

$$CaCl_2 + 7H_2O \longrightarrow CaCl_2 \cdot 7H_2O$$

$$\text{Percentage of C} = \frac{\text{Increase in weight of KOH tube} \times 12 \times 100}{\text{Weight of coal sample taken} \times 44}$$

and $$\text{Percentage of H} = \frac{\text{Increase in weight of CaCl tube} \times 2 \times 100}{\text{Weight of coal sample taken} \times 18}$$

2. Nitrogen. About 1 gram of accurately weighed powdered coal is heated with concentrated H_2SO_4 along with K_2SO_4 (catalyst) in a long-necked flask called (Kjeldahl's flask). After the solution becomes clear, it is treated with excess of KOH and the liberated ammonia is distilled over and absorbed in a known volume of standard acid solution.The unused acid is then determined by back titration with standard NaOH solution. From the volume of acid used by ammonia liberated, the percentage of N in coal is calculated as follows.

$$\text{Percentage of N} = \frac{\text{Volume of acid used} \times \text{Normality} \times 1.4}{\text{Weight of coal taken}}$$

3. Sulphur. Sulphur is determined from the washings obtained from the known mass of coal, used in a bomb calorimeter for determination of a calorific value. During this determination, 'S' is converted into sulphate. The washings are treated with Barium chloride solution when Barium sulphate is precipitated. This precipitate is filtered, washed and heated to constant weight.

$$\text{Percentage of S} = \frac{\text{Weight of } BaSO_4 \text{ obtained} \times 32 \times 100}{\text{Weight of coal sample taken in bomb} \times 233}$$

4. Ash. Ash determination is carried out as in proximate analysis.

5. Oxygen. It is obtained by difference

$$\text{Percentage of Oxygen} = 100 - \text{percentage of } (C + H + S + N + \text{ash})$$

Importance of ultimate analysis:

(i) Carbon and Hydrogen. Greater the percentage of carbon and hydrogen better is the coal in quality and calorific value. However, hydrogen is mostly associated with the volatile matter and hence, it affects the use to which the coal is put.

***(ii)* Nitrogen.** Nitrogen has no calorific value, and hence its presence in coal undesirable. Thus, a good quality coal should have very little nitrogen content.

***(iii)* Sulphur.** Sulphur although contributes to the heating value of coal, yet on combustion products like acids (SO_2 and SO_3) which have harmful effects of corroding the equipments and also cause atmospheric pollution. Sulphur is usually, present to the extent of 0.5 to 3.0% and derived from ores like iron pyrites, gypsum etc., and mines along with the coal.

Presence of sulphur is highly undesirable in coal to be used for making coke for iron industry, since it is transferred to the iron metal and badly affects the quality and properties of steel. Moreover, oxides of sulphur pollute the atmosphere and leads to corrosion.

***(iv)* Oxygen.** Oxygen content decreases the calorific value of coal. High oxygen content coal is characterized by high inherent moisture, low calorific value, and low coking power. Moreover oxygen is in combined form with hydrogen in coal and thus, hydrogen available for combustion is lesser than actual one. An increase in 1% oxygen content decreases the calorific value by about 1.7% and hence, oxygen is undesirable. Thus, a good quality coal should have low percentage of oxygen.

SOLVED PROBLEMS BASED ON ANALYSIS

Problem. *A sample of coal weighing 1.232 g was heated at 105°C for an hour and the residue weighed 1.186 g. The residue was then heated in a muffle furnace at 940°C for seven minutes and the residue now weighed 1.126 g. Finally the residue was heated in presence of air at 750°C till a constant weight (0.080 g) was attained. Calculate the results and explain what conclusions can you draw.* (SET-4, Jan 2012)

Solution. Given data

Sample weight – 1.232 g

Residue weight = 1.186 g

After heating residue weight = 1.126 g

Mass of moisture in coal sample =1.232 – 1.186 = 0.046 g

Mass of volatile matter = 1.186 – 1.126 = 0.060 g

$$\text{Percentage of moisture} = \frac{0.046 \times 100}{1.232} = 3.733\%$$

$$\text{Percentage of volatile matter} = \frac{0.060 \times 100}{1.232} = 4.870\%$$

$$\text{Percentage of ash} = \frac{0.080 \times 100}{1.232} = 6.494\%$$

$$\text{Percentage of fixed 'C'} = 100 - (3.73 + 4.87 - 6.49)$$

$$= 100 - 2.11 = 97.89\%$$

5.6 CALORIFIC VALUE

"When a unit mass (or volume) of the fuel is burnt completely, the total quantity of heat liberated is known as *calorific value*".

Units of heat:

1. **Calorie.** "Calorie" is the amount of heat required to raise the temperature of one gram of water through one degree centigrade (15–16°C).
2. **Kilocalorie.** "Kilocalorie" is equal to 1000 calories. This is the unit of metric system and may be defined as "the quantity of heat required to raise the temperature of one kilogram of water through one degree centigrade" thus, 1 kcal = 1000 cal.
3. **British Thermal Unit (B.Th.U.).** B.Th.U. is defined as "The quantity of heat required to raise the temperature of one pound of water through one degree Fahrenheit (60–61°F). This is the English System Unit.

$$1 \text{ B.Th.U.} = 252 \text{ cal} = 0.252 \text{ kcal}$$

$$\therefore \quad 1 \text{ kcal} = 3.968 \text{ B.Th.U.}$$

4. **Centigrade Heat Unit (C.H.U.).** C.H.U. is "the quantity of heat required to raise the temperature of 1 pound of water through one degree centigrade." Thus,

$$1 \text{ kcal} = 3.968 \text{ B.Th.U.} = 2.2 \text{ C.H.U.}$$

Higher or Gross Calorific Value (HCV or GCV). "The total amount of heat produced; when unit mass of the fuel has been burnt completely and the products of combustion have been cooled to room temperature is known as ***Higher or Gross calorific value***".

Usually, all fuels contain some hydrogen and when the calorific value of hydrogen-containing fuel is determined experimentally, the hydrogen is converted into steam. If the products of combustion are condensed to the room temperature (15°C or 60°F), the latent heat of condensation of steam also gets included in the measured heat, which is then called "***Higher or Gross calorific value***".

Lower or Net Calorific Value (LCV or NCV). "The net heat produced, when unit mass (or volume) of the fuel is burnt completely and the products are permitted to escape is known as ***Lower or Net calorific value.***"

In actual use of any fuel, the water vapour and moisture etc., are not condensed and escape as such along with hot combustion gases. Hence, a lesser amount of heat is available. Alternatively, Net or Lower calorific value (LCV)

$$\text{LCV} = \text{HCV–Latent heat of water vapour formed}$$

$$\text{LCV} = \text{HCV–Mass of hydrogen} \times 9 \times \text{Latent heat of steam}$$

Because 1 part by mass of hydrogen produces 9 parts by mass of water. The latent heat of steam is 587 kcal/kg or 1060 B.Th.U./lb of water vapour formed at room temperature (15°C or 60°F).

Units of calorific value. The calorific value is generally expressed in calorie/gram (cal/g) or kilocalorie/kg (kcal/kg) or British Thermal Unit/lb (B.Th.U./lb) in case of solid or liquid fuel.

In case of gaseous fuels the units used are
Kilocalorie/cubic metre (kcal/m^3) or (B.Th.U./ft^3)

$$1 \text{ kcal/kg} = 1.8 \times \text{B.Th.U./lb}$$

$$1 \text{ kcal/M}^3 = 0.1077 \times \text{B.Th.U./ft}^3$$

$$1 \text{ B.Th.U./ft}^3 = 9.3 \text{ kcal/m}^3$$

5.6.1 Determination of Calorific Value (Bomb Calorimeter)

This apparatus is used to find the calorific value of solid and liquid fuels.

Construction. A simple sketch of bomb calorimeter is shown in Fig. 5.2. It consists of a strong cylindrical stainless steel bomb in which the combustion of fuel is made to take place. The bomb has a lid, which can be screwed to the body of bomb so as to make a perfect gas-tight seal. The lid is provided with two stainless steel electrodes and an oxygen inlet valve. To one of the electrodes, a small ring is attached. In this ring a Nickel or stainless steel crucible can be supported. The bomb is placed in a copper calorimeter which is surrounded by an air-jacket to prevent heat loss due to radiation. The calorimeter is provided with an electrically operated stirrer and Beckmann's thermometer, which can be read accurately temperature difference up to $\frac{1}{100}$ th of a degree.

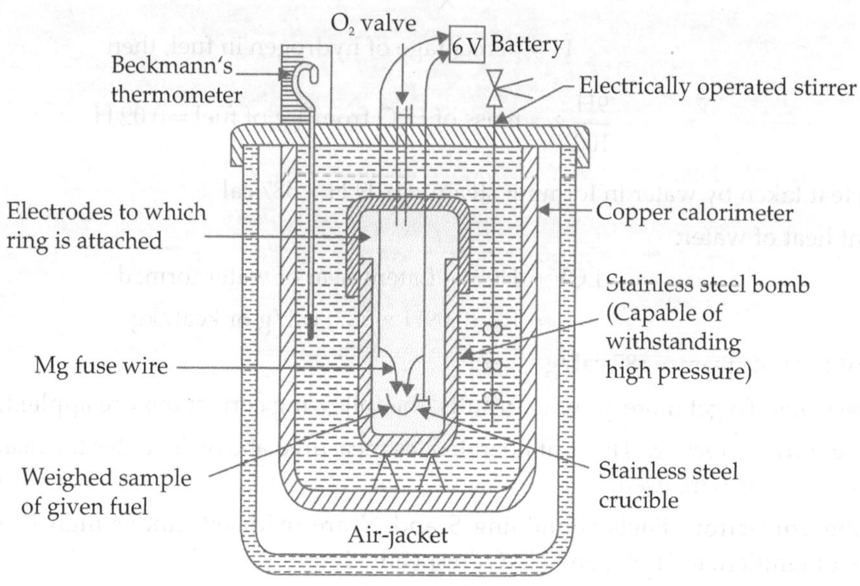

FIGURE 5.2 **Bomb calorimeter.**

Working. A known mass (about 0.5 to 1.0 g) of the given fuel is taken in clean crucible. The crucible is then supported over the ring. A fine magnesium wire, touching the fuel sample, is then stretched across the electrodes. The bomb lid is tightly screwed and bomb filled with oxygen to 25 atm pressure. The bomb is then, lowered into copper calorimeter, containing a known mass of water. The stirrer is operated and initial temperature of the water is noted. The electrodes are then connected to 6 volt battery and circuit is completed. The sample burns and heat is liberated. Uniform stirring of water is continued and the maximum temperature attained is recorded.

Calculation:

Let X = mass in gram of fuel sample taken in crucible

W = mass of water in the calorimeter

w = water equivalent in gram of calorimeter, stirrer, thermometer, bomb etc.

t_1 = initial temperature of water in calorimeter

t_2 = final temperature of water in calorimeter

L = higher calorific value in fuel in cal/g

Heat liberated by burning of fuel = XL

And heat absorbed by water and apparatus etc. = $(W + w)(t_2 - t_1)$

But heat liberated by the fuel = Heat absorbed by water, apparatus etc.

$$XL = (W + w)(t_2 - t_1)$$

$$\text{HCV of a fuel (L)} = \frac{(W + w)\,t - t}{X} \text{ cal/g or kcal/kg}$$

If H = percentage of hydrogen in fuel, then

$$\frac{9H}{100}\text{ g} = \text{mass of } H_2O \text{ from 1 g of fuel} = 0.09\text{ H}$$

∴ Heat taken by water in forming steam = 0.09 H × 587 cal

Latent heat of water:

$$\text{LCV} = \text{HCV} - \text{Latent heat of water formed}$$

$$= L - (0.09H \times 587) \text{ cal/g or kcal/kg}$$

Latent heat of steam = 587 cal/g

Corrections. To get more accurate results, the following corrections are applied,

1. Fuse wire correction. The heat liberated, as measured above, includes the heat given out by ignition of the fuse wire used.

2. Acid correction. Fuels containing S and N are oxidised, under high pressure and temperature of ignition, to H_2SO_4 and HNO_3 respectively

$$S + 2H + 2O_2 \longrightarrow H_2SO_4 + \text{heat}$$

$$2N + 2H + 3O_2 \longrightarrow 2HNO_3 + \text{heat}$$

Formations of these acids are exothermic reactions. So, the measured heat also includes the heat given out during the acid formation.

The amount of these acids is analysed from washings of bomb by titration, while H_2SO_4 alone is determined by precipitation as $BaSO_4$. The correction for 1 mg of S is 2.25 cal, while for 1 ml of N/10 HNO_3 formed is 1.43 cal.

3. Cooling correction. Time taken to cool the water in calorimeter from maximum temperature to room temperature is noted. From the rate of cooling (dt°/minute) and the actual time taken for cooling (t minutes), the cooling correction of $dt \times t$ is added to the rise in temperature.

$$L = \frac{(W + w)(t_2 - t_1 + \text{cooling correction}) - [\text{Acid} + \text{Fuse correction}]}{\text{Mass of fuel (X)}}$$

SOLVED PROBLEMS ON CALORIFIC VALUE

Problem 1. *Calculate the gross and net calorific values of coal having the following composition: Carbon: 87%, Hydrogen: 5%, Sulphur: 0.5%, Nitrogen: 0.5%, Oxygen: 3%, Ash: 4%; latent heat of steam: 587 cal/gram.* (SET-1, Jan 2012)

Solution. Given data

Carbon: 87%, Hydrogen: 5%, Sulphur: 0.5%, Nitrogen: 0.5%, Oxygen: 3%, Ash: 4%; latent heat of steam: 587 cal/gram.

$$\text{Dulong's formula GCV} = \frac{1}{100}\left[8080 \times C + 34500\left(H - \frac{O}{8}\right) + 2240 \times S\right]$$

$$GCV = \frac{1}{100}\left[8080 \times 87 + 34500\left(5 - \frac{3}{8}\right) + 2240 \times 0.5\right]$$

$$GCV = \frac{1}{100}\left[702960 + 34500 \times \frac{37}{8} + 1120\right]$$

$$GCV = \frac{1}{100}[702960 + 159562.5 + 1120]$$

$$GCV = \frac{1}{100}[863642.5]$$

$$GCV = 8636.425 \text{ kcal/kg}$$

$$LCV = [GCV - 0.09\, H \times 587] \text{ kcal/kg}$$

$$LCV = [8636.425 - 0.09 \times 5 \times 587]$$

$$LCV = [8636.425 - 264.15]$$

$$LCV = 8372.275 \text{ kcal/kg}$$

Problem 2. *Calculate the gross and net calorific values of coal having the following composition: Carbon: 85%, Hydrogen: 8%, Sulphur: 1%, Nitrogen: 2%, Oxygen: 2%, Ash: 2%; latent heat of steam: 587 cal/gram.*

(SET-3, Jan 2012)

Solution. Given data

Carbon: 85%, Hydrogen: 8%, Sulphur: 1%, Nitrogen: 2%, Oxygen: 2%, Ash: 2%; latent heat of steam: 587 cal/gram

$$\text{Dulong's formula GCV} = \frac{1}{100}\left[8080 \times C + 34500\left(H - \frac{O}{8}\right) + 2240 \times S\right]$$

$$GCV = \frac{1}{100}\left[8080 \times 85 + 34500\left(8 - \frac{2}{8}\right) + 2240 \times 1\right]$$

$$GCV = \frac{1}{100}\left[686800 + 34500 \times \frac{31}{4} + 2240\right]$$

$$GCV = \frac{1}{100}[686800 + 267375 + 2240]$$

$$GCV = \frac{1}{100}[956415]$$

$$GCV = 9564.15 \text{ kcal/kg}$$

$$LCV = [GCV - 0.09 \text{ H} \times 587] \text{ kcal/kg}$$

$$LCV = [9564.15 - 0.09 \times 8 \times 587]$$

$$LCV = [9564.15 - 422.64]$$

$$LCV = 9141.51 \text{ kcal/kg}$$

Problem 3. *A bomb calorimeter whose water equivalent is 144 g, was filled with 1458 g of water. When 0.945 g of coal was burnt in it the temperature of water rise by 5.95°C, calculate the HCV of the fuel. If the hydrogen percentage in the fuel was 7, calculate LCV of the fuel (Latent heat of condensation of steam is 587 cal/g).* (SET-4, Feb 2011, SET-2, Jan 2012)

Solution. Given data

Weight of fuel burnt X = 0.945 g

Water in calorimeter W = 1458 g

Water equivalent weight of calorimeter w = 144 g

Initial temperature t_1 = 13.25°C

Final temperature t_2 = 19.20°C

$$\text{HCV of fuel (L)} = \frac{(W + w)(t_2 - t_1)}{X} \text{ kcal/kg}$$

$$\text{HCV of fuel (L)} = \frac{(1458 + 144)(19.2 - 13.25)}{0.945} \text{ kcal/kg}$$

$$\text{HCV of fuel (L)} = \frac{(1602)(5.95)}{0.945} \text{ kcal/kg}$$

$$\text{HCV of fuel (L)} = 10086.66 \text{ kcal/kg}$$

5.7 PETROLEUM OR CRUDE OIL

Petroleum or crude oil (petra = rock;oleum = oil), is a dark greenish-brown, viscous oil found deep in earth's crust. It is composed mainly of various hydrocarbons (like straight-chain paraffins, cycloparaffins or naphthalenes, olefins and aromatics), together with small amounts of organic compounds containing oxygen, nitrogen and sulphur. The oil is, usually, found floating upon a layer of brine and has a layer of gas on top of it. The average composition of crude petroleum is: C = 79.5 to 87.1%; H = 11.5 to 14.8%; S = 0.1 to 3.5%; N + O = 0.1 to 0.5%.

5.7.1 Classification of Petroleum

The chemical nature of the crude petroleum varies with the part of the world in which it is found. There appears, however, to be three principle varieties:

1. Paraffinic-base type crude is mainly composed of the saturated hydrocarbons from CH_4 to $C_{35}H_{72}$ and a little of the naphthalenes and aromatics. The hydrocarbons from $C_{18}H_{38}$ to $C_{35}H_{72}$ are semi-solids, called 'waxes'.
2. Asphaltic-base type crude contains mainly cycloparaffins or naphthalenes with smaller amount of paraffin's and aromatic hydrocarbons.
3. Mixed-base type crude contains both paraffinic and asphaltic hydrocarbons and are, generally, rich in semi-solid waxes.

5.7.2 Origin of Petroleum

Theories based on origin of petroleum:

1. Carbide theory (Inorganic theory). Petroleum is originated from inorganic source. Molten metals react with coal deposits to form carbides which on further reactions such as hydrogenation, isomerization, and polymerization alkylation, etc., form different hydrocarbons.

$$Ca + 2C \longrightarrow CaC_2 \qquad CaC_2 + 2H_2O \longrightarrow Ca(OH)_2 + C_2H_2$$

$$4Al + 3C \longrightarrow Al_4C_3 \qquad Al_4C_3 + 12H_2O \longrightarrow 4Al(OH)_3 + 3CH_4$$

$$Mg + 2C \longrightarrow MgC_2 \qquad MgC_2 + 2H_2O \longrightarrow Mg(OH)_2 + C_2H_2$$

$$3\,HC \equiv CH \xrightarrow{Cat}$$ 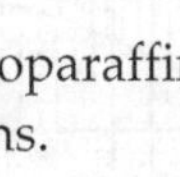

2. Organic or Engler's theory. Sea animals killed due to volcanic eruption were buried under the earth by earthquake, which were decayed and decomposed to form petroleum. Engler obtained petroleum like liquid by decomposition of fish land animal fat at high temperature and pressure. This theory explains the presence of N and S. Sea water and fossils in petroleum but not the presence of chlorophyll, high resin content and coal deposits in the petroleum.

3. Modern theory. Petroleum is formed by decay and decomposition of biological material from both plants and animals by radioactive elements or bacteria. This theory explains the presence of material like brine, coal, resins and optically active substances of plant and animal origin in the petroleum.

5.7.3 Mining of Petroleum

It is done by drilling holes in the earth's crust and sinking pipes upto the oil-bearing porous rocks. Oil, usually, Pushes out itself due to hydrostatic pressure of natural gas. Alternatively, it may be mechanically pumped up by using either lift pump or air-lift-pump. The latter consists of two co-axial pipes, lowered into the base of oil bed. Compressed air is forced through the outer pipe, whereby oil comes out through the inner pipe. The oil is conveyed to refinery by a system of pipelines.

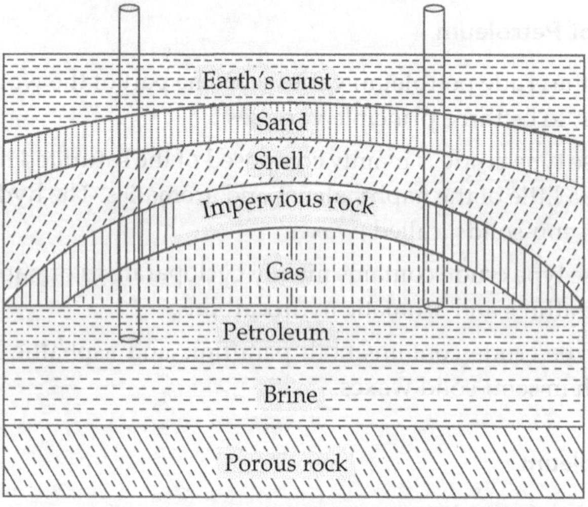

FIGURE 5.3 **Mining of ptroleum.**

Refining of crude oil. The crude oil is separated into various useful fractions by fractional distillation and finally converted into desired specific products. The process is called "refining of crude oil" and the plants set up for the purpose, are called the oil refineries. The process of refining involves the following steps:

Step 1. Separation of water (Cottrell's process). The crude oil from the oil well is an extremely stable emulsion of oil and salt water. The process of freeing oil from water consists in allowing the crude to flow between two highly charged electrodes. The colloidal water-droplets coalesce to form large drops, which separate out from the oil.

Step 2. Removal of harmful sulphur compounds. It involves in treating oil with copper oxide. A reaction occurs with sulphur compounds, which results in the formation of copper sulphide (a solid), which is then removed by filtration.

Step 3. Fractional distillation. The crude oil is then heated to about 400°C in an ion retort, whereby all volatile constituents, except the residue (alphalt or coke) are evaporated. The hot vapours are then passed up a "fractionating column", which a tall cylindrical tower containing a number of horizontal stainless steel trays at short distances. Each tray is provided with small chimney, covered with a loose cap. As the vapours go up, they become gradually cooler and fractional condensation takes place at different heights of column. Higher boiling fractions condense first; while the lower boiling fractions turn-by-turn. Various principal fractionation products thus obtained are given in Table 5.1.

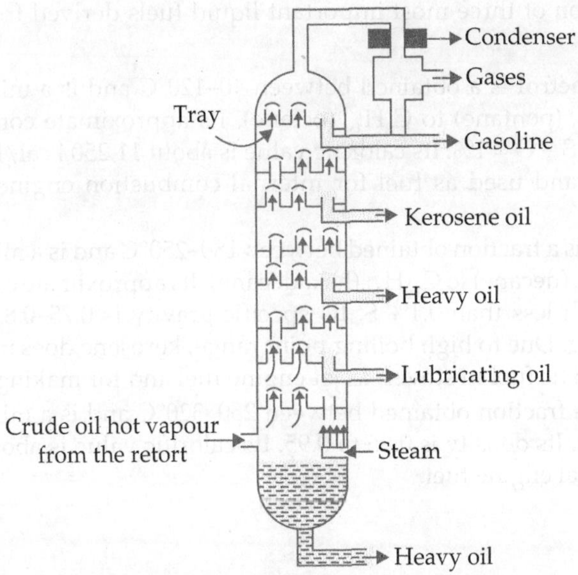

FIGURE 5.4 **Refining of petroleum.**

Table 5.1

Fraction	*Boiling range (°C)*	*Approximate composition*	*Uses*
Uncondensed gases	Up to room temperature	$C_1 - C_4$	Fuel gases: Refrigerants; production of carbon black, hydrogen; synthesis of organic chemicals.
Crude naphtha on refractionation yields,	30 – 150°C	$C_5 - C_{10}$	
(*i*) Petroleum ether (*ii*) Petrol or gasoline (*iii*) Benzene derivatives	30 – 70°C 70 – 120°C 120 – 150°C	$C_5 - C_6$ $C_6 - C_8$ $C_8 - C_{10}$	Solvent Motor fuel; dry cleaning; petrol gas. Solvent; dry cleaning
Kerosene oil Heavy oil	150 – 250°C 250 – 400°C	$C_{11} - C_{16}$ $C_{15} - C_{18}$	Fuel; illuminant; oil gas As fuel for diesel engines; converted to gasoline by cracking.
Refractionation gives,			
(*i*) Gas oil, (*ii*) Fuel oil, (*iii*) Diesel oil			
Residual oil on fractionation by vacuum distillation gives,	Above 400	$C_{17} - C_{40}$	
(*i*) Lubricating oil (*ii*) Paraffin wax (*iii*) Vaseline (*iv*) Pitch		$C_{17} - C_{20}$ $C_{20} - C_{30}$ $C_{20} - C_{30}$ $C_{30} - C_{40}$	Lubrication Candles; boot polish; wax paper; etc. Toilets; ointments; lubrication. Paints, road surfacing
Petroleum coke (on redistilling tar)			As fuel.

A brief description of three most important liquid fuels derived from petroleum is given below:

(*i*) **Gasoline or petrol** is a obtained between 40–120°C and is a mixture of hydrocarbons such as C_5H_{12} (pentane) to C_8H_{18} (octane). Its approximate composition is: C = 84%, H = 15%, N + S + O = 1%. Its calorific value is about 11,250 kcal/kg. It is highly volatile, inflammable and used as fuel for internal combustion engines of automobiles and aeroplanes.

(*ii*) **Kerosene oil** is a fraction obtained between 180–250°C and is a mixture of hydrocarbons such as $C_{10}H_{22}$ (decane) to $C_{16}H_{34}$ (hexadecane). Its approximate composition is: C = 84%, H = 16%; with less than 0.1% S. Its specific gravity is 0.75–0.85. Its calorific value is 11.100 kcal/kg. Due to high boiling point range, kerosene does not vaporize easily. It is used as domestic fuel in stoves, as jet engine fuel and for making oil gas.

(*iii*) **Diesel oil** is a fraction obtained between 250–320°C and is a mixture $C_{15}H_{32}$ to $C_{18}H_{38}$ hydrocarbons. Its density is 0.86 to 0.95. Its calorific value is about 11,000 kcal/kg. It is used as a diesel engine fuel.

5.8 CRACKING

Cracking is defined as "the decomposition of bigger hydrocarbon molecules into simpler, low boiling hydrocarbons of lower molecular weight".

For example, $$\underset{\text{b.p.} = 174°\text{C}}{C_{10}H_{22}} \xrightarrow{\text{Cracking}} \underset{(n\text{-pentane})}{C_5H_{12}} + \underset{(\text{pentane})}{C_5H_{10}}$$

b.p. = 36°C

Of all the fractions obtained by fractionation of petroleum, gasoline has the largest demand as a motor fuel, but the yield of this fraction is only 20% of the crude. Also the quality of so called 'straight-run' gasoline is not high. It has to be properly blended. Moreover, there is a surplus of heavier petroleum fractions. To overcome these difficulties, the middle and heavy fractions are cracked to get petrol. The petrol made by cracking has far better characteristics (as for as the internal combustion engine is connected) that 'straight-run' petrol. There are two methods of cracking in use.

5.8.1 Thermal Cracking

The heavy oils are subjected to high temperature and pressure; when the bigger hydrocarbon molecules break down to give smaller molecules of the paraffins, olefins plus some hydrogen. This process may be carried out either in 'liquid-phase' or 'vapour-phase'.

(*a*) **Liquid-phase thermal cracking.** The heavy oil or gas oil stock is cracked at a suitable temperature of 475–530°C and under pressure of 100 kg/cm^2. The cracked products are then separated in a fractionating column. The yield is 50–60% and octane rating of the petrol produced is 65–70.

(*b*) **Vapour-phase thermal cracking.** The cracking oil is first vaporized and then cracked at about 600–650°C and under a low pressure of 10–20 kg/cm^2. This process is suitable only for those oils, which may be readily vaporized. It requires less time than the liquid-phase method. Petrol obtained from vapour-phase cracking has better anti-knock properties, but poorer stability than petrol from liquid-phase cracking.

5.8.2 Catalytic Cracking

The quality and yield of gasoline produced by cracking can be greatly improved by using a suitable catalyst like aluminum silicate, $Al_2(SiO_3)_3$ or alumina, Al_2O_3.

Advantages:

1. The yield of petrol is higher.
2. The quality of petrol produced is better.
3. A much lower pressure (about 1–5 kg/cm^2) is needed in catalytic cracking.
4. The cracking process can easily be controlled, so the desired products can be obtained.
5. The evolution of by-product gases can be minimized; thereby the yield of desired petrol is higher.
6. The product of cracking contains a higher amount of aromatic and hence, it possess better anti-knock characteristics.

There are two methods of catalytic cracking in use:

(*a*) **Fixed-bed catalytic cracking.** The oil vapours are heated in a pre-heater to cracking temperatures (420–450°) and then forced through a catalytic chamber (containing artificial clay mixed with zirconium oxide) maintained at 425–450°C and 1.5 kg/cm^2 pressure. During their passage through the tower, about 40% of the charge is converted into gasoline and about 2.4% carbon is formed. The latter gets absorbed on the catalyst bed. The vapours produced are then passed through a fractionating column, where heavy oil fractions condense. The vapours are then led through a cooler, where some of the gases are condensed along with gasoline and uncondensed gases move on. The gasoline containing some dissolved gases is then sent to a 'stabilizer', where the dissolved gases are removed and pure gasoline is obtained.

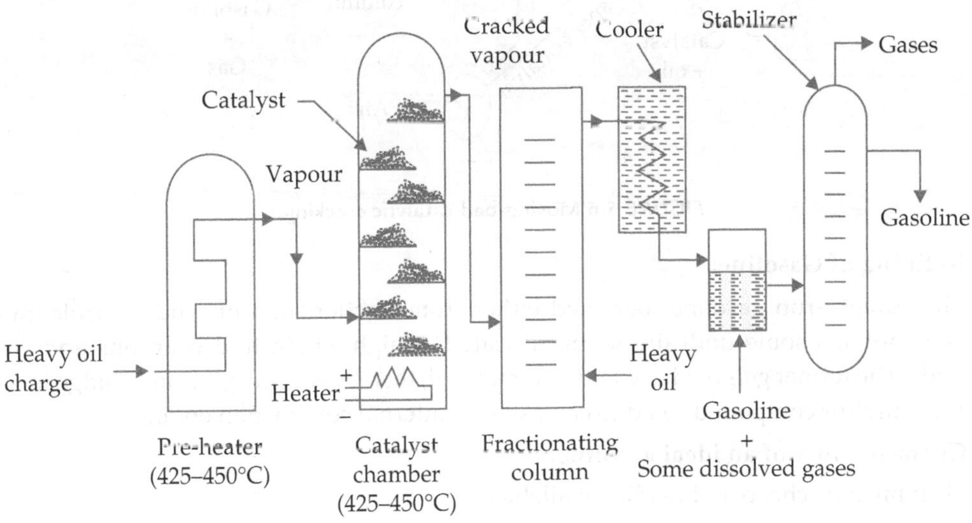

FIGURE 5.5 **Fixed-bed catalytic cracking.**

The catalyst, after 8 to 10 hours, stops functioning, due to the deposition of black layer of carbon, formed during cracking. This is reactivated by burning off the deposited carbon. During the reactivation interval, the vapours are diverted through another catalyst chamber.

(*b*) **Moving-bed catalytic cracking.** The solid catalyst is very finely powdered, so that it behaves almost as a fluid, which can be circulated in gas stream. The vapours of cracking stock (gas oil, heavy oil, etc.) mixed with fluidized catalyst is forced up into a large reactor 'bed' in which cracking of the heavier into lighter molecules occurs. Near the top of the reactor, there is a centrifugal separator (called cyclone), which allows only the cracked oil vapours to pass on to the 'fractionating column', but retains all the catalyst powder in the reactor itself. The catalyst powder gradually becomes heavier, due to coating with carbon, and settles to the bottom, from where it is forced by an air blast to regenerator (maintained at 600°C). In regenerator, carbon is burnt and the regenerated catalyst then flows through a stand-pipe for mixing with fresh batch of incoming cracking oil. At the top of the regenerator, there is a separator, which permits only gases (CO_2, etc.) to pass out, but holds back catalyst particles.

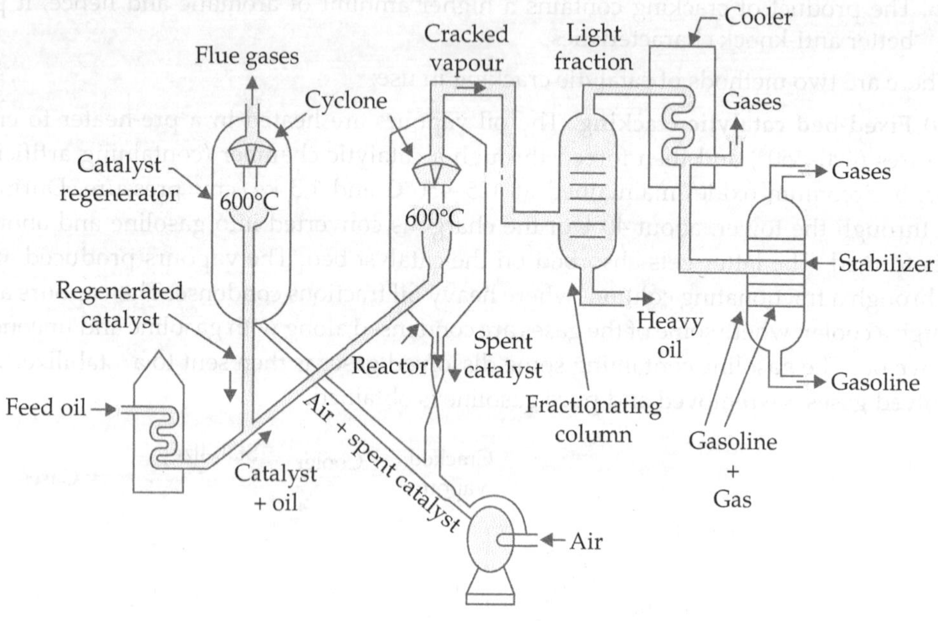

FIGURE 5.6 **Moving-bed catalytic cracking.**

Refining of Gasoline:

The straight-run gasoline (obtained either from fractionation of crude petroleum or by synthesis) contains some undesirable unsaturated straight-chain hydrocarbons and sulphur compounds. The former gets oxidized and polymerized, thereby causing gum and sludge formation on storing. Sulphur compounds lead to corrosion of internal combustion engine.

Characteristics of an ideal gasoline:

1. It must be cheap and readily available.
2. It must burn clean and produce no corrosion, etc., on combustion.
3. It should mix readily with air and afford uniform manifold distribution, *i.e.*, it should easily vaporize.
4. It must be knock-resistant.
5. It should not pre-ignite easily.

6. It should not tend to decrease the volumetric efficiency of the engine.
7. It must have a high calorific value.

Refining:

1. The sulphur compounds are generally removed by treating gasoline with an alkaline solution of sodium plumbite with controlled addition of S. This refining process (called 'sweetening') converts sulphur compounds into disulphides (Doctor's process).

$$\underset{}{2RSH} + \underset{\text{Sodium plumbite}}{Na_2PbO_2} \longrightarrow Pb(SR)_2 + 2NaOH$$

$$Pb(SR)_2 + S \longrightarrow PbS(s) + \underset{\text{Disulphide}}{RSSR}$$

The PbS(s) is removed by filtration. Since the disulphides in gasoline are also objectionable, so the disulphides so-farmed are extracted with a suitable solvent.

2. Olefines and colouring matter of gasoline are, usually, removed by percolating gasoline through 'fuller's earth', which adsorbs preferentially only the colours and olefins.
3. After the refining of gasoline, some inhibitors are added. They retard the oxidation reactions, thereby improving storing qualities of gasoline.

Even after refining, the gasoline does not have good combustion qualities and it is, generally, blended suitable with other fractions produced by catalytic cracking processes to meet the required specifications.

Reforming. Reforming is a process of bringing about structural modifications in the components of straight run gasoline (prepared by the fractional distillation of the crude oil), with the primary object of improving its anti-knock characteristics. It is carried out either thermally or in the presence of a catalyst.

1. Thermal reforming. It is carried out in a reactor (or heater) at 500 – 600°C and a pressure of 85 atm. The feed-stocks straight-run gasoline. To avoid the formation of gas (at the expense of gasoline), the conditions are controlled carefully by quenching (rapid cooling) the products by spray of cold oil. The reformed products are then fractionated to remove residual gas (stabilizing).

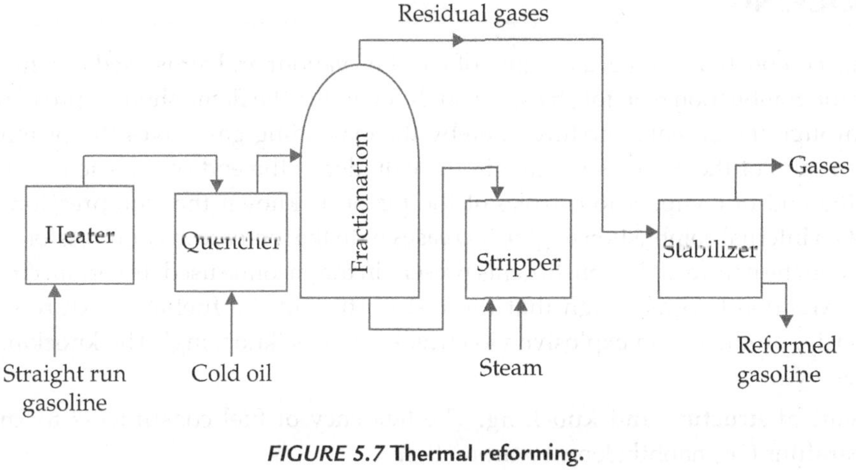

FIGURE 5.7 **Thermal reforming.**

During thermal reforming, some cracking also occurs to yield alkanes and alkenes; and they might undergo dehydrogenation, followed by cyclization dehydrocyclization to yield naphthalene's (aromatization). Conversion of *n*-alkanes to branched-chain alkanes also takes place during thermal reforming.

2. Catalytic reforming. To get better and yield of gasoline's, catalytic reforming [Pt (0.75%) supported on alumina] is carried out by using either a fixed-bed or fluidized-bed at 460 – 530°C and 35 – 50 atmospheric pressure. The main reactions during catalytic reforming process are:

Dehydrogenation:

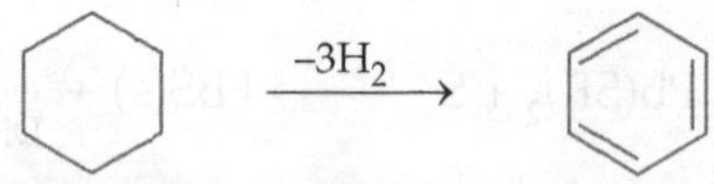

Cyclohexane Benzene

Hydrocracking:

$$\underset{n\text{-decane}}{CH_3(CH_2)_8CH_3} \xrightarrow{H_2/Pt} \underset{n\text{-pentane}}{2CH_3(CH_2)_3CH_3}$$

Isomerization:

$$\underset{n\text{-hexane}}{CH_3(CH_2)_4CH_3} \longrightarrow \underset{\text{2-methylpentane}}{\underset{\displaystyle CH_3}{CH_3\overset{}{C}HCH_2CH_2CH_3}}$$

The hydrogen thus-evolved during catalytic reforming is used to:

(*a*) hydrogenate alkenes,

(*b*) promote cracking of large alkane molecules (hydrocracking), and

(*c*) remove sulphur in the feed-stock as H_2S gas.

5.9 KNOCKING

In an internal combustion engine, a mixture of gasoline vapour and air is used as a fuel. After the initiation of the combustion reaction, by spark in the cylinder, the flame should spread rapidly and smoothly through the gaseous mixture; thereby the expanding gas drives the piston down the cylinder. The ratio of the gaseous volume in the cylinder at the end of the suction-stroke to the volume at the end of compression-stroke of the piston is known the 'compression ratio'. The efficiency of an internal combustion engine increases with the increase in compression ratio, which is dependent on the nature of the constituents present in the gasoline used. In certain circumstances the rate of oxidation becomes high that the last portion of the fuel-air mixture gets ignited instantaneously, producing an explosive violence, known as 'knocking'. The knocking results in loss of efficiency.

Chemical structure and knocking. The tendency of fuel constituents to knock in the following paraffins (*i.e.*, naphthalenes) > aromatics.

Thus, olefins of the same carbon-chain length possess better anti-knock properties than the corresponding paraffins and so on.

Octane number (introduced by Edger in 1872). It has been found that *n*-heptane knocks very badly and hence, its anti-knock value has arbitrarily been given zero. On the other hand, isooctane (2, 2, 4-trimethyl pentane), gives very little knocking, so its anti-knock value has been given as 100. Thus, octane number (or rating) of a gasoline (or any other internal combustion engine fuel) is the percentage of iso-octane in a mixture iso-octane and *n*-heptane, which matches the fuel under test in knocking characteristics. In this way, an '80-octane' fuel is one which has the same combustion characteristics as an 80 : 20 mixture of iso-octane and *n*-heptane.

n-heptane
octane number = 0

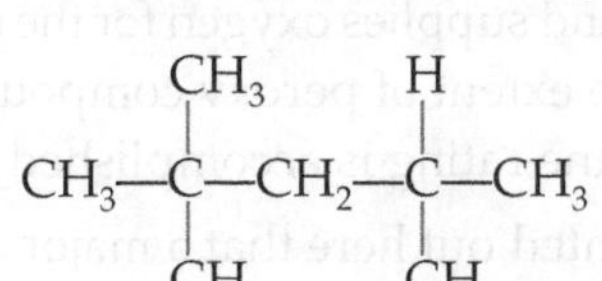

Iso-octane
octane number = 100

Cetane number. Quality of diesel is expressed in terms of cetane number. Hexadecane ($C_{16}H_{34}$) (or) cetane which ignites rapidly and 1-methylnaphthalene which ignites slowly are taken as standard and the igniting property of the fuel under study is compared with that of the mixture of the two to find its cetane number.

n-hexadecane
cetane number = 100

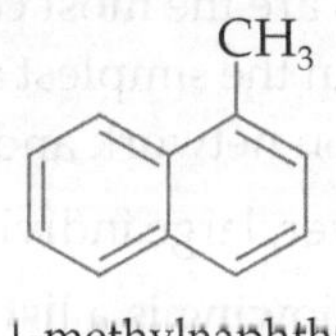

1-methylnaphthalene
cetane number = 0

5.10 ANTI-KNOCKING AGENTS

Improvement of Anti-knock Characteristics of a Fuel. The octane number of I.C. fuels can be raised by the addition of such extremely poisonous materials as tetraethyl lead, $(C_2H_5)_4Pb$ or TEL and diethyl telluride, $(C_2H_5)_2Te$. In motor spirit (or motor fuel), about 0.5 ml and in aviation fuels, about 1 ml of TEL is added per litre of petrol.

According to the most accepted theory, TEL is converted into a cloud of finely divided lead oxide particles in the cylinder and these particles react with any hydrocarbon peroxide molecules formed, thereby slowing down the chain oxidation reaction and thus, decreasing the chances of any early detonation. However, deposit of lead oxide is harmful to the engine life. Consequently, in order to help the simultaneous elimination of lead oxide formed from the engine, a small amount of ethylene dibromide is also added to petrol. Ethylene dibromide removes lead oxide as volatile lead bromide along with the exhaust gases. The presence of sulphur compounds in petrol reduces the effectiveness of the TEL.

$$Pb + \begin{array}{c} CH_2Br \\ | \\ CH_2Br \end{array} \longrightarrow PbBr_2 + CH_2{=}CH_2$$

5.11 UNLEADED PETROL

Octane rating of petrol is in leaded petrol is increased by adding tetra ethyl lead or tetra methyl lead. However, combustion of lead petrol leads to formation of litharge (PbO), which deposition the inner wall of cylinder and jams the piston. Also leaded petrol cannot be used in automobiles equipped cost catalytic converter, because lead present in exhaust gas poisons the catalyst, thereby destroying the active sites. Alternative method of increasing octane number of petrol is to add high octane compounds like iso-pentane, iso-octane, ethyl benzene, iso-propyl benzene, methyl tertiary butyl ether (MTBE). Out of these MTBE is preferred, because it contains oxygen in the form of ether group and supplies oxygen for the combustion of petrol in internal combustion engines, thereby reducing the extent of peroxy compound formation. Unleaded petrol is one where in the enhancement of octane rating is accomplished without the addition of lead compounds.

It may be pointed out here that a major advantages of unleaded petrol is that it permits the use of catalytic converter attached to the exhaust in automobiles. A catalytic converter contains a catalyst (rhodium), which converts the toxic gases (CO and NO) to harmless gases (CO_2 and N_2 respectively). Moreover, it oxidizes unburnt hydrocarbon into CO_2 and H_2O.

5.12 GASEOUS FUELS

Gaseous fuels are the most convenient because they require the least amount of handling charges and are used in the simplest and most maintenance-free burner systems. Gas is delivered "on tap" via distribution network and so is suited for areas with a high population or industrial density.

However, large individual consumers do have gasholders and some produce their own gas.

The following is a list of the types of gaseous fuel:

- Fuels naturally found in nature:
  - ➢ Natural gas
  - ➢ Methane from coal mines
- Fuel gases made from solid fuel:
  - ➢ Gases derived from coal
  - ➢ Gases derived from waste and biomass
  - ➢ From other industrial processes (blast furnace gas)
- Gases made from petroleum:
  - ➢ Liquefied Petroleum gas (LPG)
  - ➢ Refinery gases
- Gases from oil gasification
- Gases from some fermentation process

Gaseous fuels in common use are liquefied petroleum gases (LPG), Natural gas, producer gas, blast furnace gas, coke oven gas etc. The calorific value of gaseous fuel is expressed in kilocalories per normal cubic metre (kcal/Nm^3) *i.e.*, at normal temperature (20°C) and pressure (760 mm Hg).

Properties of gaseous fuels. Since most of gas combustion appliances cannot utilize the heat content of the water vapour, gross calorific value is of little interest. Fuel should be compared based on the net calorific value. This is especially true for natural gas, since increased hydrogen content results in high water formation during combustion.

5.13 NATURAL GAS

Natural gas is obtained from wells dug in the oil-bearing regions. When natural gas occurs along with petroleum in oil-wells, it is called wet gas. The wet gas is treated to remove propane, propene, butane and butene, which are used as LPG. On the other hand, when the gas is associated with crude oil, it is called dry gas.

Before use the natural gas is purified to remove objectionable ingredients such as water, dust, grit, H_2S, CO_2, N_2 and heavier liquefiable hydrocarbons (propane, butane, butane, etc.).

The approximate composition of natural gas is CH_4 = 70–90%, C_2H_6 = 5–10%, H_2 = 3%, $CO + CO_2$ = rest. The calorific value varies from 12,000 to 14,000 kcal/m^3.

Uses:

1. It is an excellent domestic fuel and can be conveyed over very large distances in pipelines.
2. It has recently been used in the manufacture of a number of chemicals by synthetic processes.
3. It is also used as raw material for the manufacture of carbon black (a filler for rubber) and hydrogen (used in ammonia synthesis).
4. Synthetic proteins are obtained by microbiological fermentation of methane.

Natural gas is mainly divided into two types:

1. Liquefied Petroleum Gas (LPG)
2. Compressed Natural Gas (CNG)

5.13.1 Liquefied Petroleum Gas (LPG)

Liquefied Petroleum Gas (LPG) or bottled gas or refinery gas is obtained as a byproduct during the cracking of heavy oils or from the natural gas.

When used for vehicles, LPG is a mixture of propane and butane (this is called autogas). When compressed, it turns into liquid form. 1 litre of LPG liquid is equivalent to 270 litre of LPG vapour. The main constituents of LPG are *n*-butane, isobutene, butylenes and propane with little or no propylene and ethane.

LPG is dehydrated, desulphurised and traces of odorous organic sulphides (mercaptans) are added to give warning of gas leak. LPG is supplied under pressure in containers under the trade name like Indane,Bharat gas, etc. Its calorific value is about 27,800 kcal/m^3.

Uses. The largest use of LPG as domestic fuel and industrial fuel. However, there is an increasing trend to use LPG as motor fuel.

Advantages:

- LPG is cheaper than petrol (up to 50%).
- It produces less exhaust emissions than petrol.
- It is better for the engine and it can prolong engine life.
- In some vehicles, it can provide better performance.
- It has a higher octane rating than petrol (108 compared to 91).

Disadvantages:

- It isn't highly available.
- The initial cost for converting your vehicle to LPG can cost up to $ 3000. However the average car can repay the cost of the conversion in about 2 years.
- It has a lower energy density than petrol.
- No new passenger cars come readily fitted with LPG (they have to be converted).
- The gas tank takes up a considerable amount of space in the car boot.

Liquid LPG (autogas with 60% propane and 40% butane) has an energy density of about 26.8 MJ/L.

LPG is not as available as petrol and diesel, but can be found at 45% of service stations in Australia (there are 3200 outlets). LPG in Australia is mainly refined locally.

Burning 100 L of LPG emits about 160 kg of carbon dioxide into the atmosphere.

5.13.2 Compressed Natural Gas (CNG)

Natural gas is compressed to high pressure of about 1,000 atmospheres is known as CNG. A steel cylinder containing 15 kg of CNG contains about 2×10^4 L or 20 m^3 of natural gas at 1 atmospheric pressure. CNG is now being used as a substitute for petrol and diesel, since it is comparatively much less pollution causing fuel. During its combustion, no sulphur and nitrogen gases are evolved. Moreover, no carbon particles are ejected. Hence it is better fuel than petrol/diesel for automobiles. However, initial cost of engine designed to use CNG as a fuel is comparatively higher than that of engine designed to use petrol/diesel. In Delhi, it is mandatory for all buses, taxis and auto to use CNG as fuel.

Why is CNG preferred over LPG:

1. CNG is a much safer fuel, since it ignites at higher temperature than petrol and diesel.
2. The conversion of gasoline operated automobiles into CNG operated vehicle is very easy.
3. The operating cost of CNG fuel is much lower compared to gasoline operation.
4. Combustion of CNG leads to lesser CO emissions than gasoline.
5. CNG mixes better with air than liquid fuels.
6. Emission from CNG operated vehicle contain no unregulated pollutants (*e.g.*, smoke, SO_2, SO_3, C_6H_6, HCHO).

5.14 COMBUSTION

Combustion is the process of oxidation of the fuel by oxygen and subsequent liberation of energies like heat, light etc.

All the fuels contain carbon which contributes the major portion of heat evolved.

Example. Combustion of carbon in oxygen

$$C + O_2 \longrightarrow CO_2 + 97 \text{ kcal}$$

For proper combustion, the substance must be brought to its ignition temperature

Ignition temperature. The minimum temperature at which the substance ignites and burns without further addition of heat from outside.

Sulphur and hydrogen too get oxidized to SO_2 and H_2O but the heat produced is unimportant due to it being small.

The air which is mostly employed for the combustion of the fuels contain 21% of oxygen by volume and 23% of oxygen by weight, the remaining being nitrogen.

The gaseous products of combustion are mainly CO, CO_2, N_2, SO_2, H_2O, known as flue gases.

While studying the combustion of fuels C, H, S, CO and hydrocarbons should be considered for complete oxidation in excess of air.

Calculation. To find out the amount of oxygen and hence the amount of air required for the combustion of a unit quantity of fuel, it is necessary to apply some elementary principles as outlined below:

1. Substance always combines in definite proportions and these proportions are determined by the molecular masses of the substances involved and the products formed.

   Example. $C + O_2 \longrightarrow CO_2$

   Mass proportions 12 32 44

   This equation indicates that mass proportions of C, O and CO_2 formed are 12 : 32 : 44
2. 22.4 litres of any gas at STP(0°C and 760 mm of Hg pressure) has a mass equal to its mole. Thus 22.4 litre of CO_2 at STP will have a mass of 44 g, because 44 g is its molar mass.
3. Air contains 21% of O_2 by volume and 23% of O_2 by mass. This means that 1 kg of O_2 is supplied by $\frac{1 \times 100}{23}$ = 4.35 kg of air

   Similarly 1 m^3 of O_2 is supplied by $\frac{1 \times 100}{21}$ = 4.76 m^3 of air.
4. Molecular mass of air is taken as 28.94 g/mol.
5. Minimum O_2 required = Theoretical O_2 required –O_2 present in the fuel.
6. Minimum O_2 required should be calculated on the basis of complete combustion. If combustion products contain CO and O_2, then excess of O_2 is found by substracting the amount of O_2 required to burn CO to CO_2.
7. The mass of the dry flue gases formed should be calculated by balancing the carbon in the fuel and carbon in the flue gases.

8. The mass of any gas can be converted to its volume at certain temperature and pressure by using the gas equation.

$$PV = nRT$$

9. Since the main elements present in most of the fuels are C, H, O and S, combustion of fuel results in the formation of gases like CO_2, H_2O, SO_2 etc.

Combustion of Carbon: 12 kg of Carbon requires 32 kg of Oxygen for complete combustion.

∴ The amount of Oxygen required for complete combustion of C kg of carbon is given as $\frac{32}{12} \times C$ kg.

Combustion of Hydrogen:

$$\underset{4}{2H_2} + \underset{32}{O_2} \longrightarrow \underset{36}{2H_2O}$$

4 kg of Hydrogen requires 32 kg of Oxygen required for complete combustion of hydrogen is given as $\frac{32}{4} \times H$ kg.

It may be pointed here that of the total amount of hydrogen, some of it is present in the combined form (*i.e.*, as H_2O). As this combined hydrogen is a non-combustible substance, so it does not take part in combustion.

Now 1 part of hydrogen combines chemically with 8 parts by mass of Oxygen.

The amount of hydrogen available for combustion

$$= \text{Mass of hydrogen} - \frac{\text{Mass of Oxygen}}{8}$$

$$= \left(H - \frac{O}{8}\right)$$

Hence, the oxygen required for combustion of hydrogen

$$= \frac{32}{4}\left(H - \frac{O}{8}\right) = 8\left(H - \frac{O}{8}\right)$$

Combustion of Sulphur:

$$\underset{32}{S} + \underset{32}{O_2} \longrightarrow \underset{64}{SO_2}$$

32 kg of sulphur requires 32 kg of oxygen for complete combustion.

∴ The amount of oxygen required for complete combustion of S kg of sulphur is given as $\frac{32}{32} \times S$ kg.

On combining the above three combustion equations, we get the equation for the theoretical oxygen requirement for the combustion of C, H and S as

$$= \frac{32}{12} \times C + 8\left[H - \frac{O}{8}\right] + S$$

where C, H, O and S are the masses of C, H, O and S for 1 kg of fuel.

Since the percentage by weight and volume of oxygen in air is 23 and 21, the amount of air required theoretically for combustion of 1 kg of fuel is given as

Minimum weight of air required for combustion

$$= \frac{100}{23}\left[\frac{32}{12}\times C + 8\left[H - \frac{O}{8}\right] + S\right] kg$$

Minimum volume of air required for combustion

$$= \frac{100}{23}\left[\frac{32}{12}\times C + 8\left[H - \frac{O}{8}\right] + S\right] m^3$$

SOLVED PROBLEMS ON COMBUSTION

Problem 1. *Calculate the weight and volume of air required for the combustion of 1 kg of carbon. Give the composition of the combustion products.*

Solution. Weight of air required for combustion of 1 kg carbon

$= \text{weight of C} \times (32/12) \times (100/23)$

$= 1 \text{ kg} \times (32/12) \times (100/23) = 11.5943 \text{ kg (or 11,594.3 g)}$

$\therefore$ Vol. of air required for combustion of 1 kg of carbon

$= \text{weight of air in g} \times (22.4 \text{ L/mol}) \times (\text{mol/mol.wt})$

$= 11{,}594 \text{ g} \times (22.4 \text{ L/mol}) \times (\text{mol}/28.94 \text{ g})$

$= 0.897 \times 10^4 \text{ L or } 8.97 \text{ m}^3 \text{ of air}$

Carbon and Oxygen are present in combustion products

The amount of CO_2 produced theoretically is

$1000 \text{ g} + 11{,}594.3 \text{ g} \longrightarrow 12594.3 \text{ g}$

Because $C + O_2 \longrightarrow CO_2$, hence 12.594 kg of CO_2 is produced.

Problem 2. *The percentage composition of a sample of coal is found to contain C = 76%, H_2 = 5.2%, O_2 = 12.8%, N_2 = 2.7%, S = 2.1% and ash = 2.2%. Calculate the minimum (i) weight (ii) volume of air at STP, required for complete combustion of 1 kg of coal and percentage composition by weight of dry products.*

Solution. Weight of elements per kg of fuel

C = 76% = 0.76 kg

H_2 = 5.2% = 0.052 kg

O_2 = 12.8% = 0.128 kg

N_2 = 2.7% = 0.027 kg

S = 2.1% = 0.012 kg and

ash = 2.2% = 0.022 kg

Weight of oxygen needed for

$$C = 0.76 \times \frac{32}{12} = 2.026 \text{ kg}$$

$$H_2 = 0.052 \times \frac{16}{2} = 0.416 \text{ kg}$$

$$S = 0.012 \times \frac{32}{32} = 0.012 \text{ kg}$$

Total weight of O_2 needed = 2.026 + 0.416 + 0.012 = 2.454 kg

Weight of dry products of combustion for

$$C = 0.76 \times \frac{44}{12} = 2.786\ (CO_2)$$

$$N_2 = 0.027\ (N_2)$$

$$S = 0.012 \times \frac{64}{32} = 0.024\ (SO_2)$$

Weight of oxygen to be supplied = weight of oxygen needed–weight of oxygen present
= 2.454 – 0.128 = 2.326 kg

$\therefore$ Weight of air needed = $2.326 \times \frac{100}{23} = 10.113$ kg

$\therefore$ Weight of nitrogen to be supplied with air = weight of air – weight of oxygen
= 10.113 – 2.326 = 7.787 kg

Products of combustion means gaseous products of combustion but not ash.

Products of combustion, excluding water are,

CO_2 = 2.786 kg SO_2 = 0.024 kg

N_2 = 0.027 kg (from coal) + 7.787 kg (from air) = 7.814 kg

Total products of combustion = 2.786 + 0.024 + 7.814 kg
= 10.624 kg

$\therefore$ Percentage composition by weight,

$$CO_2 = \frac{2.786 \times 100}{10.624} = 26.22\%$$

$$N_2 = \frac{7.814 \times 100}{10.624} = 73.55\%$$

$$SO_2 = \frac{0.024 \times 100}{10.624} = 0.225\%$$

Problem 3. *The percentage composition of sample of coal is C = 90%, H_2 = 3.5%, O_2 = 3.0%, N_2 = 1%, S = 0.5% and ash = 2.0%. Find the weight of air required for combustion of 1 kg of coal and composition of the products of combustion by volume if 50% air is supplied.*

Solution. Given data,

1 kg of coal contains

$$C = 900\ g, \quad H_2 = 35\ g, \quad O_2 = 30\ g,$$
$$N_2 = 10\ g, \quad S = 5\ g \text{ and ash} = 20\ g.$$

$$\therefore \quad \text{Net oxygen} = \left[\frac{32}{12} \times C + \frac{16}{2} \times H\right] - \text{oxygen present in fuel}$$

$$= \left[\frac{32}{12} \times 900 + \frac{16}{2} \times 35\right] - 30$$

$$= 2400 + 280 - 30 = 2650\ g = 2.650\ kg$$

$$\text{Weight of air required} = 2.650 \times \frac{100}{23} = 11.52\ kg$$

$$\text{Weight of air required if 50\% air is supplied} = 11.52 \times \frac{50}{100} = 5.76\ kg.$$

Problem 4. *An oil analysis gave the following results C = 85%, H = 12%, and O = 3%. Find the weight of minimum air requirement for burning of 1 kg of the fuel.*

Solution. Given data,

$$C = 850\ g, H = 120\ g, O = 30\ g,$$

$$\therefore \quad \text{Net oxygen} = \left[\frac{32}{12} \times C + \frac{16}{2} \times H\right] - \text{Oxygen present in fuel}$$

$$= \left[\frac{32}{12} \times 850 + \frac{16}{2} \times 120\right] - 30$$

$$= 2266.6 + 960 - 30 = 3196.6\ g = 3.196\ kg$$

$$\text{Minimum weight of air required} = 3.196 \times \frac{100}{23} = 13.895\ kg$$

$$\therefore \quad \text{Quantity of air required} = 13.895\ kg.$$

Problem 5. *A gaseous fuel has the following percentage composition $CH_4 = 22\%$, $H_2 = 10\%$, $CO = 26\%$, $O_2 = 2\%$, the reminder being N_2. Calculate the volume of air required for the complete combustion of 100 m^3 of fuel.*

Solution. Volume of gaseous in 1 m^3 of gaseous fuel are

$$CH_4 = 0.22\ m^3, H_2 = 0.1\ m^3, CO = 0.26\ m^3 \text{ and } O_2 = 0.02\ m^3$$

Combustion reactions:

$$CH_4 + 2O_2 \longrightarrow CO_2 + 2H_2O$$

So volume of O_2 required $0.22 \times 2 = 0.44\ m^3$

$$H_2 + \frac{1}{2}O_2 \longrightarrow H_2O$$

So volume of O_2 required = 0.1 × 0.5 = 0.05 m^3

$$CO + \frac{1}{2}O_2 \longrightarrow CO_2$$

So volume of O_2 required = 0.26 × 0.5 = 0.13 m^3

Total O_2 required for the above combustion reactions are = 0.62 m^3

Net mass of O_2 required = Oxygen needed–Oxygen present in the fuel = 0.62 – 0.02 = 0.6 m^3

∴ Volume of air required for perfect combustion of 1 m^3 of gas

$$= 0.60 \times \frac{100}{21} = 2.857 \text{ m}^3$$

∴ Volume of air required for 100 m^3 of gas = 2.857 × 100 = 285.7 m^3 of gas.

EXERCISE

1. What are fuels? Explain their classification.
2. Write the differences between solid, liquid and gaseous fuels.
3. What is coal? Explain the analysis of coal:

   (*a*) Proximate analysis (*b*) Ultimate analysis.
4. Define calorific value. Explain HCV and LCV.
5. How do you determine the calorific value by bomb calorimeter?
6. What is petroleum? How it is classified?
7. Write a short note on

   (*a*) Mining of petroleum (*b*) Origin of petroleum.
8. Explain refining of petroleum.
9. Define Cracking. Explain thermal and catalytic cracking.
10. Define knocking. Explain the knocking of petrol and diesel.
11. What is natural gas? Explain about LPG and CNG.
12. What is combustion? Explain briefly about combustion.

Chapter 6 CHEMISTRY OF ADVANCED MATERIALS

6.1 NANOMATERIALS

Materials having components with in size less than 100 nm at least in one dimension are known as Nanomaterials. The prefix (nano) in the word nanomaterials means a billionth (1×10^{-9} m). Atoms are very small and the diameter of a single atom can vary from 0.1 to 0.5 nm. It deals with various structures of matter having dimensions of order of a billionth of metre.

1. **Nanoparticles.** Nanoparticles are the particles the size of which ranges from 1–50 nm. Generally they are obtained from colloids. The colloidal particles have nano crystals. A large percentage of atoms in nano crystals are present on the surface.

   Nano crystals possess electronic magnetic and optical properties. Since the nanoparticles exhibit an electronic behaviour governed by the quantum physics they are also called as quantum dots.
2. **Nanochemistry or Nanoscience.** Nanoscience is defined as the study of phenomena and manipulation of materials at atomic molecular and macromolecular scales.
3. **Nanotechnology.** Nanotechnology is defined as the design characterization production and applications of structures systems and devices by controlling size and shape at 9–10 m scale or the single–atomic level.
4. **Nanomaterials.** Nanomaterials are the materials having components with size less than 100 nm.
   - Nanomaterials in **one-dimension** are layers such as a thin films or surface coatings.
   - Nanomaterials in **two-dimensions** are tubes such as nanotubes and nanowires.
   - Nanomaterials in **three-dimensions** are particles like precipitates, colloids and quantum dots.

6.1.1 Carbon Nanotubes (CNTs)

Carbon nanotubes are allotropes of carbon with a nanostructure having a length-to-diameter ratio greater than 1,000,000. When graphite sheets rolled into a cylinder their edges joined and form carbon nanotubes. Carbon nanotubes are extended tubes of rolled graphite sheets. Nanotubes naturally align into ropes and held together by van der Waals forces. But each carbon atoms in the carbon nanotubes are linked by the covalent bond.

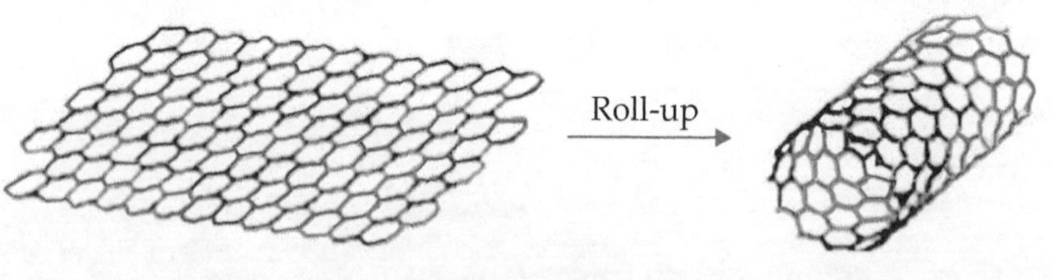

FIGURE 6.1

Structure or Types of Carbon Nanotubes. Carbon nanotubes are lattice of carbon atoms, in which each carbon is bonded to three other carbon atoms. Depending upon the way in which graphite sheets are rolled two types of CNTs are formed.

1. Single-walled nanotubes (SWNTs).
2. Multi-walled nanotubes (MWNTs).

1. Single-walled nanotubes (SWNTs). SWNTs consist of one layer of graphite. It is one-atom thick having a diameter of 2 nm and length of 100 nm. SWNTs are very important because they exhibit important electrical properties. It is an excellent conductor. Three kinds of nanotubes are resulted based on the orientation hexagon lattice.

(i) Armchair structures. The lines of hexagons are parallel to the axis of the nanotube.

(ii) Zigzag structures. The lines of carbon bonds are down the centre.

(iii) Chiral nanotubes. It exhibits twist or spiral around the nanotubes.

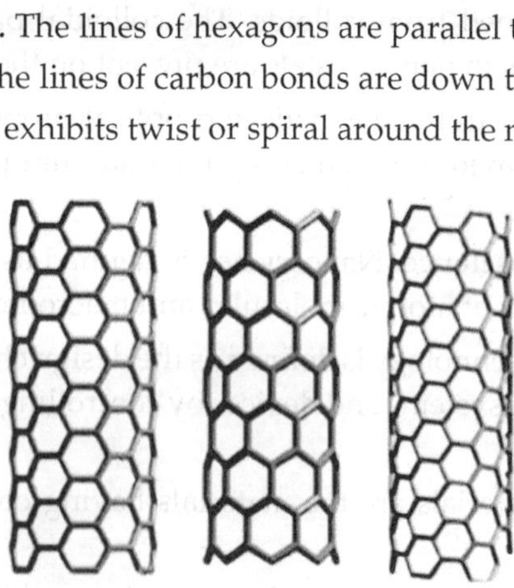

FIGURE 6.2

It has been confirmed that armchair carbon nanotubes are metallic while zigzag and chiral nanotubes are semiconducting.

2. Multi-walled nanotubes (MWNTs). MWNTs (nested nanotubes) consist of multiple layers of graphite rolled on themselves to form a tube shape. It exhibits both metallic and semiconducting properties. It is used for storing fuels such as hydrogen and methane. The inter layer distance in MWNTs is closed to the distance between graphene layer in graphite, approximately 3.3 Å.

6.1.1.1 Properties of carbon nanotubes (CNTs) or nanomaterials

Mechanical properties:

(*a*) Carbon nanotubes are very strong and their elastic flexibility is indicated by young's modulus. Young's modulus of carbon nanotubes are 10 times than that of steel.

(*b*) CNTs withstand extreme strain and tension. Most of the materials fracture on bending because of the presence of more defects but CNTs possess only few defects in the structure.

Electrical properties. The electrical properties of CNTs vary between metallic to semiconducting materials. It depends on the diameter and chirality of the nanotubes. The very high electrical conductivity of CNTs is due to the minimum defects in the structure.

Thermal conductivity. The thermal conductivity of CNTs is very high this is due to the vibration of covalent bonds. Its thermal conductivity is 10 times greater than the metal. The high thermal conductivity is due to minimum defects in the structure.

Vibrational properties. As in normal molecules the atoms in a CNT are continuously vibrating back and forth CNTs have two normal modes of vibration.

(*i*) A 1 g mode. It involves "in" and "out" oscillation.

(*ii*) E 2 g mode. It involves a squashing of the tube (oscillation between sphere and an ellipse)

The frequencies of two modes of vibrations are Raman active.

Kinctic properties. The inner tubes of MWNTs can slide without any friction within the outer nanotubes leading to an atomically perfect rotational bearing.

6.1.2 Fabrication Methods (Preparation Methods)

Various techniques are capable of synthesizing carbon nanotubes. Some of them are,

1. Carbon-Arc method.
2. Laser vaporisation method.
3. Chemical vapour deposition method (CVD method).

1. Carbon-arc method. The carbon arc discharge method initially used for producing C_{60} fullerenes is the most common and perhaps easiest way to produce carbon nanotubes as it is rather simple to undertake, however it is a technique that produces a mixture of components and requires separating nanotubes from the soot. And the catalytic metals present in crude product.

This method creates nanotubes through arc vaporisation of two carbon rods placed end to end, separated by approximately 1 nm, in an enclosure that is usually filled with inert gas (He, Ar) at low pressure (between 50 and 700 m bar). Recent investigations have shown that it is also possible to create nanotubes with the arc method in liquid nitrogen. A direct current of 50 to 100 amperes driven by approximately 20 volts creates high temperature discharge between the two electrodes. The discharge vaporises one of the carbon rods and forms a small rod shaped deposit on the other rod producing nanotubes in high yield depends on the uniformity of the plasma arc and the temperature of the deposit form on the carbon electrode.

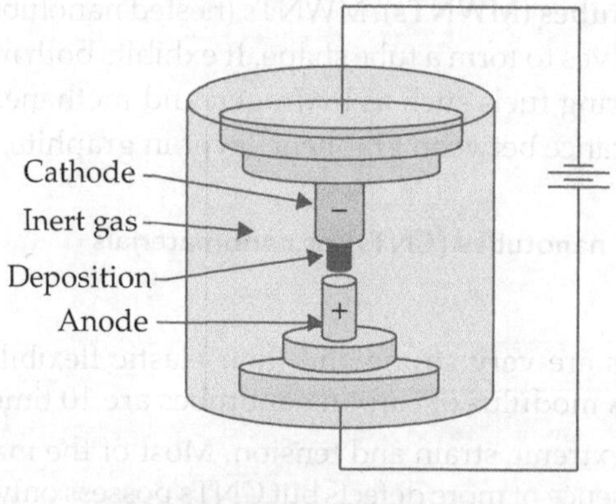

FIGURE 6.3 **Carbon-Arc method.**

2. Laser vaporisation method. This method produces single wall carbon nanotubes in high yields. A graphite target is heated to 1200°C in a quartz tube a Nd-YAG laser ablats carbon off of this graphite target. At the end of the furnace a water cooled brass cone collects the soot from the ablation. This soot containing a high percentage of single wall carbon nanotubes shown in Fig. 6.4.

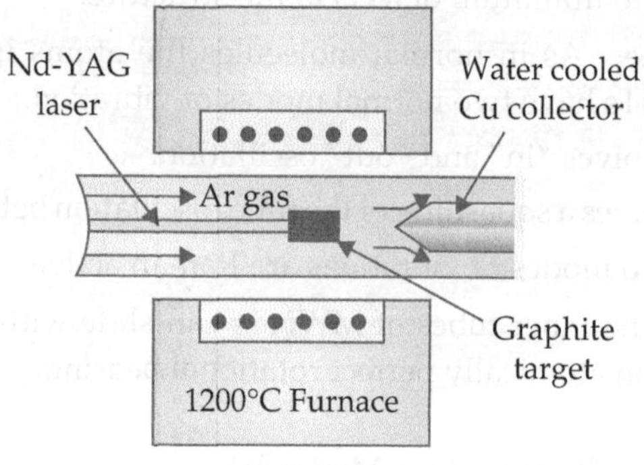

FIGURE 6.4 **Laser vaporisation method.**

3. Chemical vapour deposition (CVD) Method. Chemical vapour deposition (CVD) synthesis is achieved by putting a carbon source in the gas phase and using an energy source such as plasma or a resistively heated coil to transfer energy to a gaseous carbon molecule. Commonly used gaseous carbon sources include methane, carbon monoxide and acetylene. The energy source is used to "Crack" the molecule into reactive atomic carbon. Then the carbon diffuses towards the substrate, which is heated and coated with a catalyst (usually a first row transition metal such as Ni, Fe or Co) where it will bind. Carbon nanotubes will be formed if the proper parameters are maintained. Excellent alignment as well as positional control on nanometre scale can be achieved by using CVD. Control over the diameter, as well as the growth rate of the nanotubes can also be maintained.

Chemical vapour deposition (CVD) carbon nanotube synthesis is essentially a two-step process consisting of a catalyst preparation step followed by the actual synthesis of nanotube. The catalyst is generally prepared by sputtering a transition metal onto a substrate and then using either chemical etching or thermal annealing to introduce catalyst particle nucleation. Thermal annealing results in cluster formation on the substrate, from which the nanotubes will grow.

Ammonia may be used as the etchant. The temperatures for the synthesis of nanotubes by CVD are generally within the 650–900°C range. Typical yields for CVD are approximately 30%.

6.1.3 Fullerenes

A third newly discovered allotrope of carbon (after graphite and diamond) is Buckminster fullerene named in honour of the American architect Buckminster fuller, the designer of geodesic dome. The shape of C_{60} resembles with the domes designed by fuller. Fullerenes were discovered as an unexpected surprise during laser spectroscopy experiments at Rice University in september in 1985.

Types of fullerenes. Fullerenes are a family of carbon allotropes, molecules entirely composed of carbon in the form of hallow, sphere, ellipsoid, tube or plane. Thus fullerenes are of the following types.

(*i*) ***Spherical fullerenes.*** They looks like a soccer ball and are often called bucky balls.

(*ii*) ***Cylindrical fullerenes.*** These are called carbon nanotubes or bucky tubes.

(*iii*) ***Planar fullerenes.*** Graphene is an example of planar fullerene sheet.

Preparation. Fullerenes are prepared by vaporizing a graphite rod in a helium atmosphere mixtures of fullerenes like C_{60}, C_{70}, etc., are formed which are separated by solvent extraction. C_{60} is isolated from this mixture by column chromatography using alumina/hexane as a solvent.

Structure. Fullerenes are a class of closed–cage carbon molecule, C_n, characteristically containing 12 pentagons and a variable number of hexagons.

C_{60} is a spherical crystal of carbon atoms with an arrangement of even number of sp^2 atoms over the surface of a closed hollow cage. Each atom is trigonally linked to its three neighbouring atoms by bonds and thus forms a polyhedral network. The C_{60} molecule has a truncated icosahedrons structure.

An icosahedron is polygon with 60 vertices and 32 faces, 12 of which are pentagonal and 20 hexagonal. A carbon atom is present at each vertex. The valences of each carbon atom are satisfied by two single and one double bond. The molecule has several resonating structure and is aromatic. The fused ring aromatic system bends around and closes to form a soccer shaped molecule (a bucky ball).

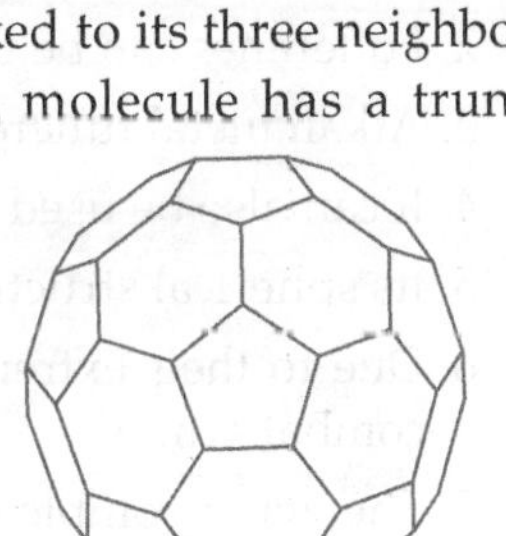

FIGURE 6.5

Other fullerenes are also been discovered. C_{70} is elongated like a rugby ball. Richard.E. Smalley of Rice University predicts that a 'Russian doll' molecule can be made by trapping C_{60} molecule inside a very large C_{240} fullerenes. Giant fullerenes with at least 600 atoms also seen to form and appear to have interesting polyhedral shapes. The smallest number of fullerene family is C_{20}.

Derivatives of fullerenes. Fullerenes readily participate in a number of chemical reactions. Its reactivity is very good; therefore, a number of fullerene derivatives can be prepared.

1. **Methylation.** Up to 24 methyl groups can be added to C_{60} polyanions.
2. **Chlorination and Bromination.** Polychlorinated and brominated derivatives have been produced.

3. Charge transfer complexes such as $[Fe(C_2H_5)_2]C_{60}$ and exohedral organometallic derivatives (Os, Pt, In) are some of the examples of fullerene derivatives.
4. **Super conducting fullerides.** C_{60} is a poor conductor of electricity. When fullerene reacts with good electron donors like alkali metals, its conductivity increases. Alkali fullerene compounds such as K_3C_{60} are superconducting. This compound has critical temperature (TC) 19.2 K. The potassium atoms occupy the octahedral sites in the cubic cell. Some other fullerides are $PbCs_2C_{60}$, Rb_3C_{60} and Ca_5C_{60}.
5. **Magnetic donors.** Organic electron donors like tetrakis dimethyl amino ethylene (TDAE) reduces C_{60} and form solid $(TDAE)C_{60}$. It is ferromagnetic solid with a curie temperature of 16 K.

Properties. The bucky ball has certain unique properties:

1. It exists as a discrete molecules.
2. C_{60} is a mustard coloured solid when the thickness of the film increases it appears brown to black.
3. It dissolves in benzene forming a deep magenta solution.
4. It is very tough. It can be accelerated to 1500 miles per hour.
5. Highest tensile strength of any known 2D structure.
6. Highest packing density.
7. It is thermally stable upto 600°C. It undergoes sublimation under vacuum at 600°C.

Engineering applications. The fullerenes derivative have been found to exhibit amazing conducting, magnetic, optical and mechanical properties.

1. Fullerenes can easily accept electrons they may be used as charge carrier in batteries.
2. Fullerenes can be used as organic photovoltaics.
3. Alkali metal fullerenes are superconductors.
4. It can also be used as soft ferromagnet.
5. Its spherical structure makes it suitable for the used as lubricant.
6. Due to their extremely resilient and nature of fullerenes are being consider for using combat armor.
7. The water soluble derivatives of fullerenes inhibit the HIV-I and are therefore usefully in fighting the HIV that leads to AIDS.

6.1.4 Engineering Applications of Nanomaterials

1. In electronics and optoelectronics:

1. Nano phosphors for affordable high definition television and flat panel displays.
2. Electroluminescent non crystalline silicon, opening the way for optoelectronic chips and possibly new type of colour displays.
3. Efficient light emitting diodes based on QD with a voltage controlled, tunable output colour.
4. Powder or plastic layers using nanoparticles as an active scattering medium.
5. Optical switches and fibres based on non-linear behaviour.

6. Transparent conducting layers.
7. Three-dimensional optical memories.

Electronics:

1. Materials for the next-generation computer chips.
2. Single-electon tunneling transistors using nanoparticles as QD.
3. Electrically conducting nanoceramics.
4. Efficient electrical contacts for semiconductor devices.

Magnetic applications:

1. Magnetic memories based on material with a high coercively.
2. Magnetoresistive materials, important for shielding components and devices.
3. Soft magnetic alloys such as finement.
4. Resistors and voltage dependent resistors.
5. High temperature superconductors using nanoparticles for flux pinning.

In optics:

1. Graded refractive index optics: special plastic lenses.
2. Anti-fogging coating for spectacles and car windows.
3. Inexpensive coloured glasses and optical filters.

2. In energy storage:

1. Novel solar cells, such as the gratzel cell based on TiO_2 materials.
2. Window layers in classical solar cells utilizing the increased band gap due.
3. High energy density (rechargeable) batteries.
4. Smart windows based on the photochrome effect or electrical magnetic orientations effects.
5. Nanocrystalline hydrogen storage materials.
6. Better thermal or electrical insulation materials, again using the higher gap.
7. Magnetic refrigerators from super paramagnetic materials.

3. In gas sensing devices:

1. Gas sensors for NO_x, SO_x, CO, CO_2, CH_4 and aromatic hydrocarbons.
2. UV sensors and robust optical sensors based on nanostructured silicon carbide (SiC).
3. Smoke detectors.
4. Ice detectors.
5. Ice detectors on aircraft wings.

Protective coatings:

1. Cost effective corrosion protection materials.
2. Elimination of pollutants in catalytic converters utilizing the large surface area of nano-materials.
3. Scratch resistance top coat using hybrid nanocomposite materials.

Medical applications:

1. Long-lasting medical implants of biocompatible nano-structured ceramic and carbides.
2. Biocompatible coating for medical applications.
3. Magnetic nanopatricles for hyperthermis.
4. Controlled drug release and drug delivery.

Catalysis:

1. Photo catalyst air and water purifiers.
2. Better activity, selectivity and life time in chemical transformations and fuel cells.
3. Precursors for a new type of catalyst.
4. Stereoselective catalyst using chiral modifiers on the surface of metal nanoparticles.

6.2 LIQUID CRYSTALS

Liquid crystals (LCs) are a state of matter that have properties between those of a conventional liquid and those of a solid crystal. For instance, an LC may flow like a liquid, but its molecules may be oriented in a crystal-like way.

The Swarm theory of BOSE. This theory was proposed by E. BOSE (1909)

1. According to this theory, since mesomorphic behaviour is shown by substances having long chains, the molecules will probably tend to arrange themselves parallel to each other.
2. In this way a number of groups or swarms are obtained. The molecules in any one swarm are parallel to one another but may not parallel to those present in another swarm.
3. The swarms may be compared to the small crystals, although the molecules in the swarm are not arranged in rigid manner and also size is not constant.
4. The turbidity of liquid crystal can be explained by swarm theory. Turbidity of liquid crystals is attributed to the scattering of light by the swarms.
5. The effect is quite similar to that observed when a clear crystalline substance or glass is in finely powdered.
6. When the temperature is raised, the size of swarms diminishes because of increased thermal movement of the molecules.
7. At the melting point when it becomes clear liquid, its size becomes too small to scatter the light.
8. The double refraction, resulting from orientation within the swarms, will disappear at the same time.

It may be noted that the mesomorphic state is not very simple. This theory has not been finally accepted as there are still many facts which remain unexplained. It is not clearly understood why certain long chain molecules which might be expected to form liquid crystals do not do so.

Uses. Liquid crystals are becoming important substances because of their electrical and optical properties and use in electronic instruments.

1. Digital displays like pocket calculators, digital wrist watches, etc.
2. Liquid crystals are also used as commercial lubricants.

3. These are most suited to biological functions. Since the colour of the liquid crystal depends sensitively on temperatures; these have been used to measure skin temperatures.
4. These are used for detecting tumors in the body by employing method called thermography.
5. These are employed as solvents for studying the structure of anisotropic molecules spectroscopically.
6. Due to their electrical and mechanical properties lying between crystalline solids and isotropic liquids these are used in gas liquid chromatography.

6.2.1 Types of Liquid Crystals

Liquid crystal is a term referring to substances that are not crystalline (solid) nor isotropic (liquid), but somewhere between the two. There are three main types, or what is scientifically known as mesophases, of liquid crystals which can be identified by their varying amounts of molecular order and positioning. This arrangement of molecules is what makes the substance either more solid or liquid.

There are many different types of liquid crystals phase, which can be distinguished by their different optical properties (such as birefringence). When viewed under a microscope using a polarized light source, different liquid crystal phases will appear to have distinct textures.

Liquid crystals can be divided into three types:

1. Thermotropic phases.
2. Lyotropic phases.
3. Metallotropic phases.

1. Thermotropic Phases. Thermotropic LCs consists of organic molecules. Thermotropic LCs exhibit a phase transition into the LC phase as temperature is changed. Thermotropic phases are those that occur in a certain temperature range. If the temperature rise is too high, thermal motion will destroy the delicate cooperative ordering of the LC phase, pushing the material into a conventional isotropic liquid phase. At too low temperature, most LC materials will form a conventional crystal. Many thermotropic LCs exhibit a variety of phases as temperature is changed.

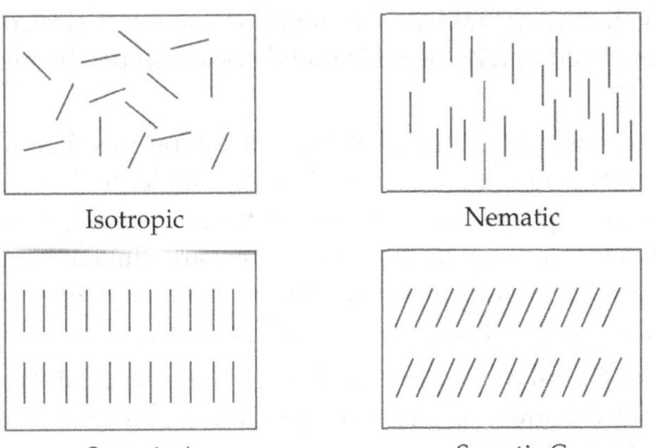

FIGURE 6.6

For instance, a particular type of liquid crystals molecule (called mesogen) may exhibit various Smectic and Nematic (and finally isotropic) phases as temperature is increased. An example of a compound displaying thermotropic LC behaviour is para-azoxyanisole.

Generally these are three types:

(*i*) Nematic phase.

(*ii*) Smectic phase.

(*iii*) Cholesteric or Chiral phase.

(*i*) Nematic phase. One of the most common liquid crystals phases is the nematic. The word *nematic* comes from the Greek (*nema*), which means "thread".

The nematic phase is the simplest form of liquid crystal and is the phase in which the crystal molecules have no orderly position and are free to move any which way. However, while they have no specific order, during this phase the molecules do tend to point in the same direction, which is what differentiates it from a pure liquid. Liquid crystal in this phase can be characterized by its thread-like appearance when looked at under a microscope. The use of nematic liquid crystal is common in telescope lenses as it allows for a clear image when researchers are confronted with atmospheric turbulence.

Most nematics are uniaxial: They have one axis that is longer and preferred, with the other two being equivalent (can be approximated as cylinders or rods). However, some liquid crystals are biaxial nematics, meaning that in addition to orienting their long axis, they also orient along a secondary axis. Nematics have fluidity similar to that of ordinary (isotropic) liquids but they can be easily aligned by an external magnetic or electric field. Aligned nematics have the optical properties of uniaxial crystals and this makes them extremely useful in liquid crystal displays (LCD).

(*ii*) Smectic phase. The smectic phase of liquid crystal, which is defined as being equivalent to the slippery, thick residue found at the bottom of soap dishes, is characterized by a slight degree of translational order in the crystal molecules which is not found in the nematic phase. While keeping similar orientation and pointing in the same direction as the molecules in nematic liquid crystal do, in this phase the molecules tend to line themselves up into layers. While these layers as a whole move freely, movement within the layers is restricted; therefore, it creates a slightly more solid substance. Smectic liquid crystal has been found to have fast electro-optical response time and because of this is used, along with nematic liquid crystal, in producing liquid crystal display (LCD) screens.

The smectic phases, which are found at lower temperatures than the nematic, form well-defined layers that can slide over one another in a manner similar to that of soap. The smectics are thus positionally ordered along one direction. In the Smectic A phase, the molecules are oriented along the layer normal, while in the Smectic C phase they are tilted away from the layer normal. These phases are liquid-like within the layers. There are many different smectic phases, all characterized by different types and degrees of positional and orientational order.

(*iii*) Cholesteric or Chiral phase. The chiral nematic phase exhibits chirality (handedness). The cholesteric phase, also known as chiral nematic phase, because it was first observed for cholesterol derivatives. Only chiral molecules (*i.e.*, those that lack inversion symmetry) can give rise to such a phase.

The chiral nematic phase is characterized by the molecules being aligned and at a slight angle to one another, stacked within very thin layers— it is the last phase before a substance becomes crystalline, or solid. This type of liquid crystal also has the characteristic of changing colour when it is exposed to different temperatures. It is for this reason that cholesteric liquid crystal is used in common household items such as thermometers and mood rings.

2. Lyotropic Phases. A lyotropic liquid crystal consists of two or more components that exhibit liquid-crystalline properties in certain concentration ranges. In the lyotropic phases, solvent molecules fill the space around the compounds to provide fluidity to the system. In contrast to thermotropic liquid crystals, these lyotropics have another degree of freedom of concentration that enables them to induce a variety of different phases.

A generic progression of phases, going from low to high amphiphile concentration, is:

1. Discontinuous cubic phase (micellar cubic phase)
2. Hexagonal phase (hexagonal columnar phase) (middle phase)
3. Lamellar Phase
4. Bicontinuous cubic phase
5. Reverse hexagonal columnar phase
6. Inverse cubic phase (Inverse micellar phase).

Even within the same phases, their self-assembled structures are tunable by the concentration: for example, in lamellar phases, the layer distances increase with the solvent volume. Since lyotropic liquid crystals rely on a subtle balance of intermolecular interactions, it is more difficult to analyze their structures and properties than those of thermo tropic liquid crystals.

3. Metallotropic Phases. Liquid crystal phases can also be based on low-melting *inorganic* phases like $ZnCl_2$ that have a structure formed of linked tetrahedra and easily form glasses. The addition of long chain soap-like molecules leads to a series of new phases that show a variety of liquid crystalline behaviour both as a function of the inorganic-organic composition ratio and of temperature. This class of materials has been named metallotropic.

6.2.2 Applications of Liquid Crystals

Liquid crystals find wide use in liquid crystal displays (LCD), which rely on the optical properties of certain liquid crystalline substances in the presence or absence of an electric field.

1. The electric field can be used to make a pixel switch between transparent or opaque on command. Colour LCD systems use the same technique, with colour filters used to generate red, green, and blue pixels. Similar principles can be used to make other liquid crystal based optical devices.
2. Thermotropic chiral LCs whose pitch varies strongly with temperature can be used as crude thermometers, since the colour of the material will change as the pitch is changed.
3. Liquid crystal colour transitions are used on many aquarium and pool thermometers as well as on thermometers for infants or baths. Other liquid crystal materials change colour when stretched or stressed.
4. Thus, liquid crystal sheets are often used in industry to look for hot spots, map heat flow, measure stress distribution patterns, and so on.

5. Liquid crystal in fluid form is used to detect electrically generated hot spots for failure analysis in the semiconductor industry.
6. Liquid crystal memory units with extensive capacity were used in space shuttle navigation equipment.
7. Many common fluids, such as soap, are in fact liquid crystals. Soap forms a variety of LC phases depending on its concentration in water.

Applications of Liquid Crystals in LCD. Liquid crystal display screens are more compact, lightweight and portable than other display screens. In addition, LCD screens require a low amount of energy and have a very long lifespan, making them more cost effective for many technologies. liquid crystal display screens are used in a wide range of products.

Television. Liquid crystal display technology is common in television screens for its light weight design and cost effectiveness. According to MSNBC, since 2007, LCD sales have surpassed that of the traditional vacuum tube televisions and offer competition to plasma television sales as well. In addition, Flat Panel TV.org states that an LCD TVs screen lifespan is upward of 60,000 hours, making it more cost effective than other TV screens. LCD televisions are available in all sizes ranging from a four inch pocket sized TV to stadium displays of more than 150 inches.

Cameras. Camera display screens often use LCD technology to save space and reduce weight. An LCD camera screen displays the digital image from the lens and allows the user to review photos stored in the camera's memory. According to IEEE, LCD is the favoured technology for digital camera screen production.

Computers. Desktop and laptop computer screens use LCD technology because of the compact design and energy efficiency. Laptop screens favour LCD technology because it requires little energy to power, extending battery life. In addition, LCD technology is lightweight, supporting the portability of the laptop unit.

Watches. Watch display screens use LCD technology to create a modern watch screen. LCD watches typically have options for customization including larger numbers and personalized backgrounds. An LCD screen also uses a low amount of energy, allowing for maximum watch battery life.

6.3 FIBER REINFORCED PLASTICS (FRP)

Fiber reinforced plastics are produced by reinforcing a plastic matrix with a high strength fiber materials such as glass, graphite, alumina, carbon, boron, beryllium and aromatic polyamide. Natural fibers such as sisal, asbestos are also used for reinforcement. Depending the desired properties of the final reinforced composite, the nature of the fiber use is decided. Fiber reinforced plastics of this fiber withstands from 400°C to – 1096°C.

Fiber reinforced plastics are characterized by

(*i*) High specific strength (*i.e.*, tensile strength/specific gravity).

(*ii*) High specific modulus (*i.e.*, modulus of elasticity/specific gravity).

The properties of fiber reinforced composites depend on

(*i*) Nature and properties of fiber and matrix materials.

(*ii*) Their relative volume fractions.

(*iii*) Orientation and distribution of fibers in matrix phase.

(*iv*) Strength of the interfacial bond between the fiber and the matrix phase.

(*v*) Length of the fibers or more appropriately aspect ratio of fibers.

Components of composite (FRP):

- **Fiber:**
  - ➢ Fibers are used to strengthen thermoplastic compounds.
  - ➢ The properties of FRP are mainly influenced by the choice of fibers.
  - ➢ There are three main types of fibers carbon, glass and Aramid fibers.
  - ➢ They have different properties.
  - ➢ All fibers have generally high stress capacity than the ordinary steel.
- **Matrix:**
  - ➢ The plastic material used in FRP is called matrix.
  - ➢ Matrix is a tough and relatively weak plastic that is reinforced by stronger fibers.
  - ➢ It transfers the forces between the fibers and protects the fiber from the environment.
  - ➢ Generally thermosetting plastics are used as matrix.
  - ➢ Vinyl esters and epoxy resins are most commonly used matrices.

Applications of FRP:

(*i*) Fiber reinforced plastics find extensive use in spacecrafts, aeroplanes, boat nulls, acid storage tanks.

(*ii*) Motor cars and building materials.

(*iii*) Melamine FRP is used for insulating and making baskets.

Advantages of FRP:

The fiber reinforced plastics have the following advantages:

(*i*) Low coefficient of thermal expansion.

(*ii*) High dimensional stability.

(*iii*) Low cost of production.

(*iv*) Good tensile strength.

(*v*) Low dielectric constant.

(*vi*) Non inflammability and corrosion and chemical resistance.

6.3.1 Types of Fiber Reinforced Plastics

Fiber Reinforced Plastics are the following three types.

6.3.1.1 Carbon fiber reinforced plastics (CFRP)

Carbon fibers can be combined usefully with a wide variety of plastic resins and polymers to yield composite materials of exceptionally high specific stiffness and strength. The most efficient reinforcing effect is obtained from continuous fibers carefully aligned in a highly oriented configuration with the proportion of fibers in any direction.

Matrix materials. Many thermosetting resins and thermoplastics are suitable for combination with carbon fibers. The thermosetting resins are epoxies or modified epoxies, polyesters, phenolics and vinyl esters are used. Thermoplastic resins are polyamides, polycarbonates, polysulphones, etc.

Properties:

1. High specific stiffness and strength.
2. High specific modulus.
3. Good mechanical strength.
4. Carbon fibers are electrically conductive along the filament axis.
5. Low thermal expansion.
6. Good chemical resistance.

Applications:

1. In structural components requiring high specific stiffness and strength.
2. Used in air frame structures, space vehicles and satellites.
3. Used in sports equipment and high speed reciprocating parts for industrial machinery.
4. Used for very high temperature machinery.

6.3.1.2 Glass fiber reinforced plastics (GFRP)

- Glass fiber is made up of sand, limestone, folic acid and other minor ingredients.
- The mixture is heated till it melts at about 1260°C.
- The molten glass is then allowed to flow through fine holes in a platinum plate.
- The glass strands are cooled, gathered and wound.
- The fibers are woven into various forms for use in composites.
- Another glass fiber prepared from aluminium + lime + borosilicate has high insulating properties, low susceptibility to moisture and high mechanical strength.

Properties:

- They have high surface area to weight. Increased surface area is susceptible to chemical attack.
- Glass fiber has good thermal insulation due to the presence of air gaps.
- Glass fiber absorbs moisture which lesser tenacity.
- Glass fiber undergoes more elongation before it breaks.
- Glass fiber is drawn at moderately low viscosity. If it is too high, fiber will break during drawing. If it is too low, the glass will form droplets.

6.3.1.3 Bullet proof plastics (Aramide Fiber Reinforced Polymer)

Aromatic poly amides are known as Aramides. The modulli of the fibers are 70–200 Gpa with ultimate elongation of 1.5–5% depending on the quality. Aramides are a family of nylons, including Kevlar and Nomex. Kevlar is the best bullet proof plastic.

Examples:

1. **Kevlar:** It is prepared by poly condensation between aromatic acid dichloride and aromatic diamines.

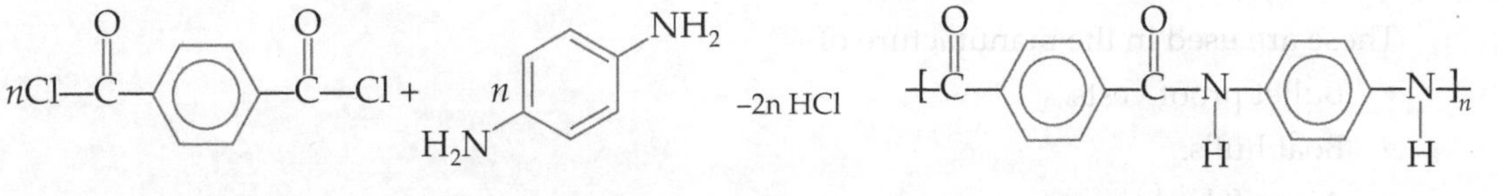

Terephthalic acid dichloride P-amino aniline Kevlar

Properties of Bullet proof plastics:

1. Kevlar is exceptionally strong (5 times stronger than steel and 10 times stronger than Aluminium).
2. It has high heat stability and flexibility.
3. The unique properties of Kevlar are due to the delocalized bonding which causes the benzene rings to be flexible.
4. Kevlar is more rigid than nylons.

Applications (Uses):

1. Kevlar is used extensively in the aerospace and aircraft industries.
2. It is used to make car parts (such as tyres, brakes, clutch linings, etc.)
3. It is used to make ropes, cables, protective clothing, bullet proof vests, motorcycle helmets and other high performance materials.

2. **Nomex:**

- This is produced by condensation reaction from the monomers *m*-phenylene diamine and isophthaloyl chloride.

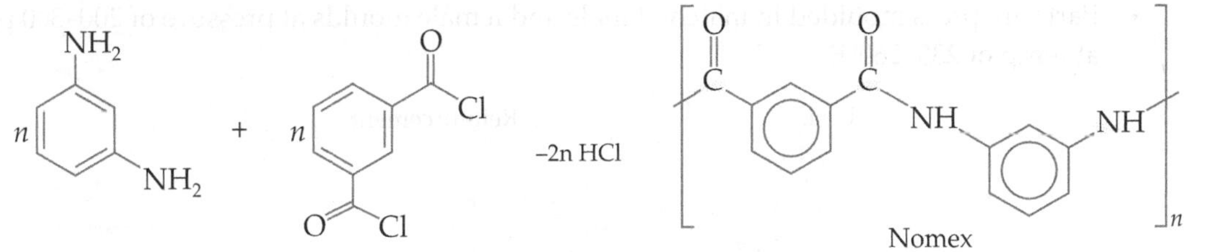

Nomex

Properties:

Fiber reinforced plastics is five times as strong as steel.

- Usually yellow in appearance.
- Low density.
- High strength.

- Thermal resistance.
- Good impact resistance.
- Good abrasion resistance.
- Good chemical resistance.
- Compressive strength.
- Some fibers degrade on exposure to UV-rays.

Uses:

These are used in the manufacture of

- Bullet proof vests.
- Boat hulls.
- Aircraft body parts.
- Clutch plates.
- Brake pads.
- Gaskets.
- Adhesive and sealants.

6.3.2 Methods for Producing FRP's

- The reinforcement may be chopped fiber, continuous filaments or woven fabrics.
- The matrix may be polyester, epoxy, phenolics, silicones, melamines, vinyl derivatives and polyamides.
  - ➢ Polyesters are widely used because of their low cost.
  - ➢ Phenolic resins are thermally stable.
  - ➢ Epoxy resins impart good mechanical strength.
  - ➢ Silicone resins impart good electrical and thermal properties.

1. Matched metal die moulding:

- This is most efficient and economical method for mass production of high strength parts.
- Parts are press moulded in matched male and female moulds at pressure of 200–300 psi at temp of 235–260°F.

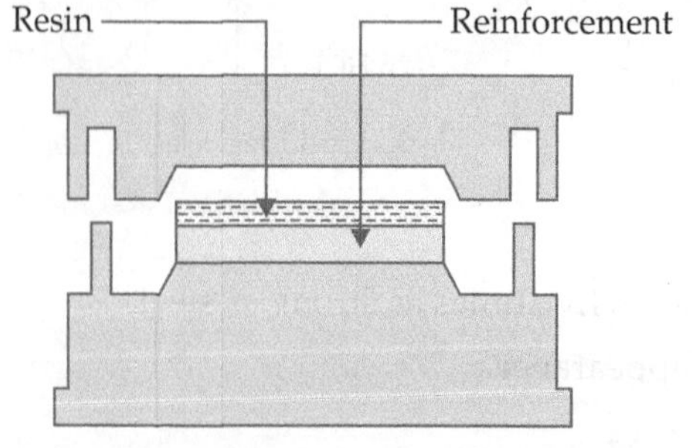

FIGURE 6.7 **Matched metal die moulding.**

2. Injection moulding:

- Pieces of fibers and matrix is mixed and passed through screw conveyor into a heated cavity of a closed matched metal mould.

3. Hand lay up:

- This is simple method for forming thermosetting plastics.
- The mould can be made of low cost wood or plaster.
- The reinforcing mat or fabric is cut to fit, laid in the mould and resin is applied by hand using a brush or roller or spray gun.
- Layers are built up to desired thickness and laminate is cured to render it hard.

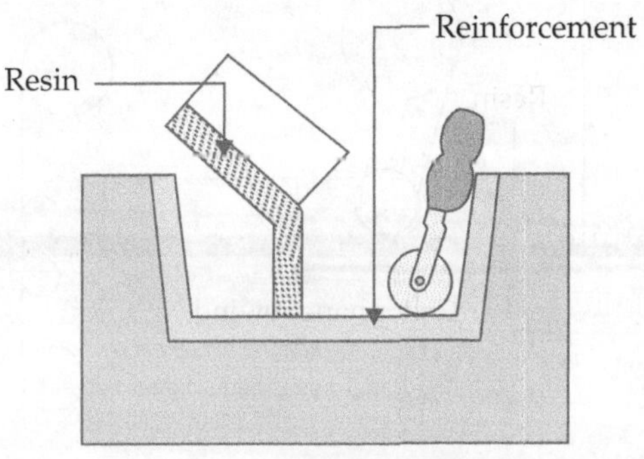

FIGURE 6.8 **Hand lay up.**

4. Spray up:

- Single mould is used in this method.
- This method is automated to some extent.
- It is used for thermosetting resins.
- Pieces of reinforcement material along with the resin are sprayed onto the surface of the mould using specially designed gun.
- Curing takes place at room temperature in presence of catalyst.

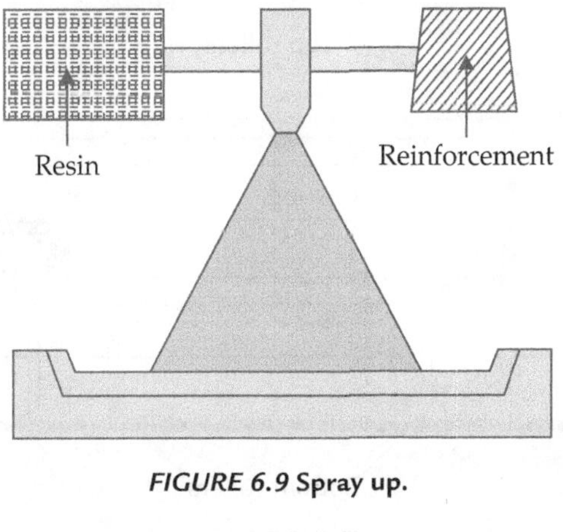

FIGURE 6.9 **Spray up.**

5. Continuous lamination:

- This is the most economical method of producing corrugated sheets.
- Reinforcing mats or fabrics are impregnated with resin, run through laminating rolls between cellophane sheets to control thickness and resin contents.
- They are then cured in a heated zone.

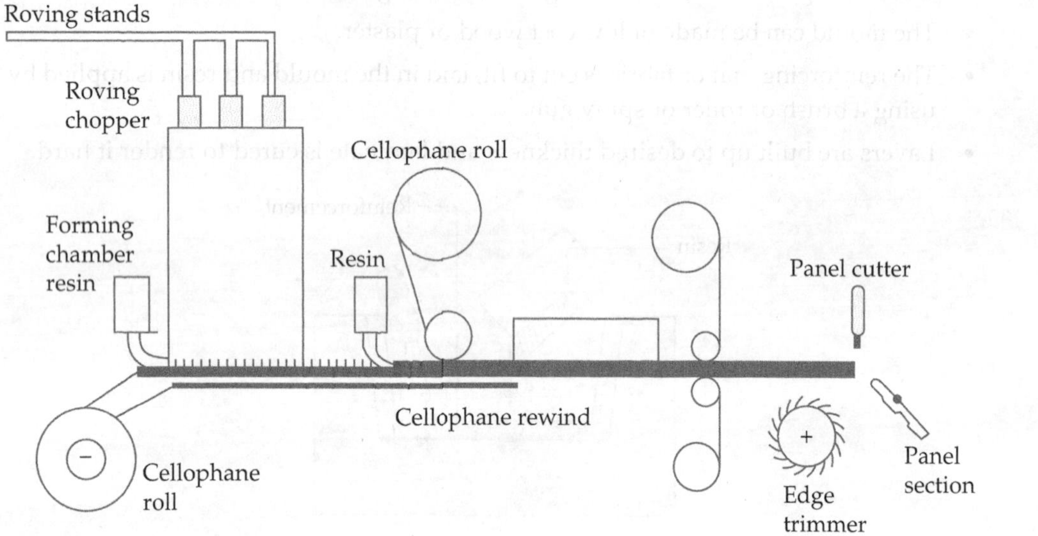

FIGURE 6.10

6. Centrifugal casting:

- This method is suitable for providing round, oval, tapered or rectangular parts.
- It offers uniform thickness and good inner and outer surfaces.
- The labour and tooling costs are low.
- In this method, chopped fibers and resins are placed inside a mandrel and are uniformly distributed as the mandrel is rotated inside an oven.

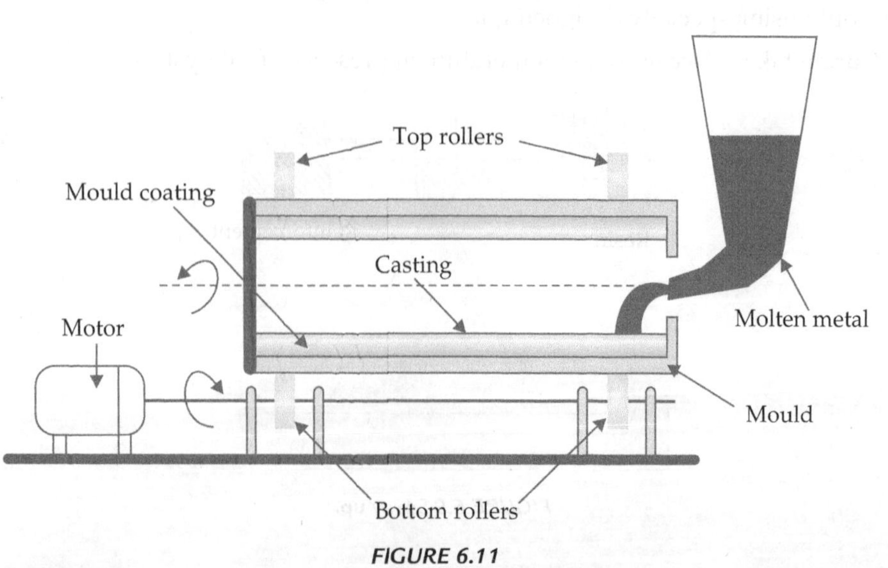

FIGURE 6.11

7. Pultrusion:

- This method is suitable for providing shapes with high unidirectional strength.
  Example. I-beams, flat stock for building siding, fishing rods, and shafts for golf clubs.
- In this continuous fiber strands combined with mat or woven fibers for cross strength, are impregnated with resin and pulled through a long heated steel die.
- The die shapes and controls the resin content.

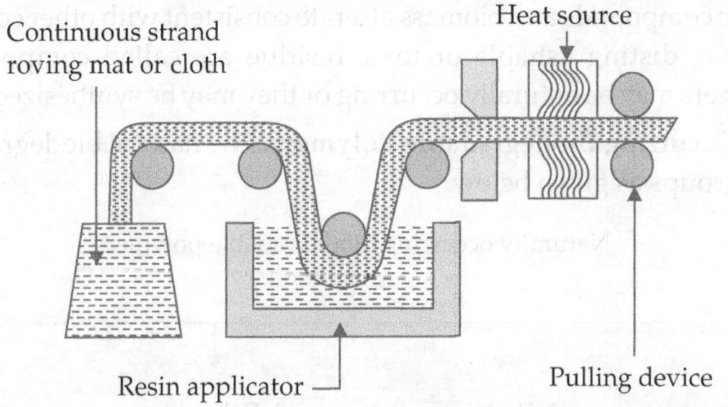

FIGURE 6.12 **Pultrusion.**

8. Filament winding:

- In this method, continuous fiber strands are wound on a suitably shaped mandrel or core and positioned in predetermined positions.
- The mandrel may be left in place permanently or removed after curing.
- Final curing is done by heating.

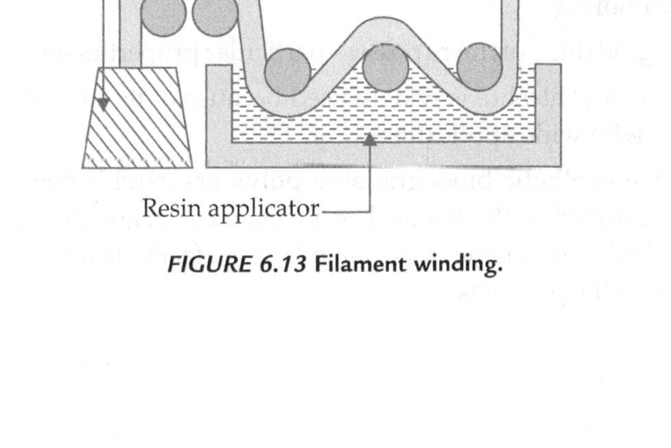

FIGURE 6.13 **Filament winding.**

6.4 BIODEGRADABLE POLYMERS

Polymers are not attacked by environmental conditions including biological attack. Polymers are degraded by oxidation, ultraviolet radiation, etc. but not by bacteria.

Biodegradable polymers are defined as a degradable polymer in which degradation results from the action of naturally occurring microorganisms such as bacteria, fungi and algae.

A plastic that undergoes degradation by biological processes during compositing to yield CO_2, water, inorganic compounds and biomass at a rate consistent with other compostable materials and leaves no visible, distinguishable or toxic residue are called compostable plastics. The biodegradable polymers may be naturally occurring or they may be synthesized by chemical means.

1. Naturally Occurring Biodegradable Polymers. The natural biodegradable polymers are classified into four groups as given below.

Naturally occurring biodegradable polymers

1. Polysaccharides *e.g.*, Starch
2. Proteins *e.g.*, Gelatin,
3. Polyesters *e.g.*, Polyhydroxy alkanoali
4. Others *e.g.*, silk, wool

FIGURE 6.14

2. Synthesised Biodegradable Polymers. There are many polymers produced from derived from petrochemical or biological sources that are biodegradable. Many of these polymers are considered as special materials for example dissolving suture material used in medical field, and biopolyesters. Because of their high prices these polymers could not gain popularity, Later a sustainable interest was created in the development of chemicals derived from farming of particular crops, which laid foundation to the synthesis of these new biodegradable polymers.

There are a number of biodegradable synthetic resins produced as list below.

(*a*) Polyalkylene esters.

(*b*) Polylactic acid and its co-polymers.

(*c*) Polyvinyl esters.

(*d*) Polyvinyl alcohol.

The above biodegradable polymers possess particular properties and potential applications.

For example polyvinyl alcohol is widely used because of its solubility in water. Polylactic acid is produced easily with wide applications.

All of the above synthetic biodegradable polymers exhibit degradation promoted by microorganisms often coupled with chemical or mechanical degradation. These polymers also contain plasticizers, antioxidants, pigments and stabilizers to facilitate their processing and enhance physical characteristics of the products.

Applications of biodegradable polymers:

1. These biodegradable polymers are synthesised from the material derived from the processing of crops or form petrochemical feed stock with normal/conventional processing methods.
2. The 'compostable bags' help in the disposal of the vegetable matter being converted to CO_2 and CH_4.
3. The problem of landfills by solid waste can be reduced.
4. Several test methods are developed to determine the compostability of environmentally degradable plastics.

Drawbacks of biodegradable polymers is that the biodegradable products at present are expensive than the products manufactured from commodity resins.

6.5 CONDUCTING POLYMERS

Most polymeric materials are poor conductors of electricity, because of the non availability of large number of free electrons in the conduction process.

"Polymeric materials have been synthesised which possess electrical conductivities on par with metallic conductors. Such polymers are known as conducting polymers".

Conducting polymers are divided into four types:

1. Intrinsically Conducting Polymer (ICP) or Conjugated π-electrons Conducting Polymer. "A polymer whose backbones or associate groups consists of delocalized electron pair or residual charge is known as **ICP**."

This type of polymers essentially contain conjugated π-electrons backbone, which is responsible for electrical charge. In an electric field, conjugated π-electrons of polymer get excited; thereby can be transported through the solid polymeric material. Overlapping of orbital's (of conjugated π-electrons) over the entire backbone results in the formation of valence bands as well as conduction bands, which extends over the entire polymer molecule. Presence of conjugated π-electron in a polymer increases its conductivity to a large extent.

Important commercially produced conducting polymers:

(*a*) Polyacetylene polymers *e.g.*, poly-*p*-phenylene, polyquinoline, polyphenylene-co-vinylene, poly-*m*-phenylenesulphide, etc.

(*b*) With condensed aromatic rings *e.g.*, polyaniline, polyanthrylene, polyphenanthrylene, etc.

(*c*) With aromatic heteroatomic and conjugated aliphatic units, *e.g.*, polypyrrole, poly thiophene, polybutadienylene, etc.

2. Doped Conducting Polymers. It is obtained by exposing a polymer to a charge transfer agent in either gas phase or in solution. The ICPs possess low conductivity ($\approx 10^{-10}$ ohm^{-1} cm^{-1}), but these possess low ionization potential and high electron affinities, so these can be easily oxidized or reduced. Consequently, the conductivity of ICP can be increased by creating either positive or negative charges on the polymer backbone by oxidation or reduction. This technique is called Doping (an analog with semiconductor). Doping is of two types:

***(i)* P-doping.** P-doping involves treating an ICP with Lewis acid, thereby oxidation process takes place and positive charges on the polymer backbone are created. Some of the P-dopant used are I_2, Br_2, AsF_5, PF_6, etc.

$$\underset{\text{Polyacetylene}}{(C_2H_2)_n} + 2FeCl_3 \xrightarrow{\text{Oxidation}} (C_2H_2)_n^+ FeCl_4^- + FeCl_2$$

$$\underset{\text{Polyacetylene}}{2(C_2H_2)_n} + 3I_2 \xrightarrow{\text{Oxidation}} 2[(C_2H_2)_n^+ I_3^-]$$

***(ii)* N-doping.** N-doping involves treating an ICP with Lewis base thereby reduction process takes place and negative charges on the polymer backbone are created. Some of the common N-dopant used are Li, Na, Ca, Tetrabutyl ammonium, $FeCl_3$, etc.

$$\underset{\text{Polyacetylene}}{(C_2H_2)_n} + \underset{\text{Lewis base}}{B} \longrightarrow (C_2H_2)_n^- B^+$$

$$\underset{\text{Polyacetylene}}{(C_2H_2)_n} + \underset{\text{Naphthyl amine}}{C_{10}H_7NH_2} \longrightarrow (C_2H_2)_n^- NH^+ + \underset{\text{Naphthalene}}{C_{10}H_8}$$

3. Extrinsically Conducting Polymers (ECP). Those polymers whose conductivity is due to the presence of "externally" added ingredient in them are known as ECPs.These are of the following two types:

***(i)* Conductive element–filled polymer.** This is a resin or polymer filled with conducting elements such as carbon black, metallic fibers, metal oxides, etc. In this, the polymer act as the binder to hold the conducting elements together in the solid entity. These polymers possess reasonably good bulk conductivity, and are generally, low in cost, light in weight, mechanically durable and strong, and easily processable in different forms, shapes and sizes.

Generally, special grade conducting carbon black is used as a filler, due to its

(*a*) Very high surface area (1000 m/g),

(*b*) High porosity, and

(*c*) Filamentous properties.

Minimum concentration of conductive filler in the polymer to start the conduction process is known as percolation threshold.

***(ii)* Blended conducting polymer.** This product is obtained by blending a conventional polymer with a conducting polymer either by physical or chemical change. Such polymers can be easily processed and possess better physical, chemical and mechanical properties.

4. Coordination Conducting Polymer (Inorganic Polymer). It is a charge transfer complex containing polymer obtained by combining a metal atom with a polydentate ligand. Although the degree of polymerization in such polymers is small (≤ 18), yet they exhibit corrosion characteristics, and are usually in doubtable in solvents.

Engineering applications:

(*i*) In rechargeable lightweight batteries based on percolate doped polyacetylene-lithium system. These are about 10 times lighter than conventional lead storage batteries. Such batteries are sufficiently flexible to fit a variety of designed configuration.

(*ii*) In optically display devices (LEDs) based on polythiophene. When the structure is electrically biased (1 to 3 V), the optical density of the film changes, *i.e.*, its colour changes. Such electrocromic systems produce coloured displays with faster switching time and better viewing than conventional LCDs.

(*iii*) In wiring in aircrafts and aerospace components.

(*iv*) In telecommunication systems.

(*v*) In anti static coatings for clothing.

(*vi*) In electromagnetic screening materials.

(*vii*) In electronic devices such as transistors and diodes.

(*viii*) In solar cells, drug delivery system for human body, etc.

6.6 SOLAR CELLS

Solar Cells (Photovoltaic cells). A solar cell is a device that converts the energy of sunlight directly into electricity by the photovoltaic effect. Sometimes the term *solar cell* is reserved for devices intended specifically to capture energy from sunlight such as solar panels and solar cells, while the term *photovoltaic cell* is used when the light source is unspecified. Assemblies of cells are used to make solar panels, solar modules, or photovoltaic arrays. Photovoltaic is the field of technology and research related to the application of solar cells in producing electricity for practical use:

1. Photons in sunlight hit the solar panel and are absorbed by semiconducting materials, such as silicon.
2. Electrons (negatively charged) are knocked loose from their atoms, allowing them to flow through the material to produce electricity. Due to the special composition of solar cells, the electrons are only allowed to move in a single direction.
3. An array of solar cells converts solar energy into a usable amount of direct current (D.C.) electricity.

Solar cells are made of thin wafers of semiconductor materials like silicon and gallium. When solar radiations fall on them, a potential difference is produced which causes flow of electrons and produces electricity. Silicon can be obtained from silica or sand, which is abundantly available and inexpensive. By using gallium arsenide, cadmium sulphide or boron, efficiency of the photovoltaic cells can be improved. The potential difference produced by a single photovoltaic cell of 4 cm^2 size about 0.4–0.5 volts and produces a current of 60 milliamperes. Figure 6.15 shows the structure of a solar cell.

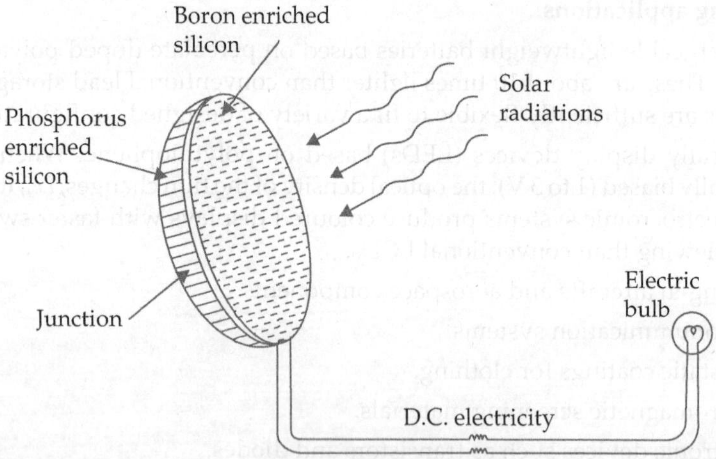

FIGURE 6.15 **Solar cell (PV cell).**

Applications of Solar Cells. Alongside a variety of consumer products—electronic watches, calculators, power for leisure equipment and tourism-there is an extensive range of applications where solar cells are already viewed as the best option for electricity supply. These applications are usually stand-alone, and exploit the following advantages of photovoltaic electricity:

- There are no fuel costs or fuel supply problems.
- The equipment can usually operate unattended.
- Solar cells are very reliable and require little maintenance.
- These are used in street-lighting, traffic signals, water pumps, etc.
- They are also used in artificial satellites for electricity generation.
- These are used for running radio and TV.
- They are used in remote areas where conventional electricity supply is a problem.

Solar Panel. Photovoltaic cells are arranged like a panel is known as solar panel.

A **Solar Panel** (**photovoltaic module** or **photovoltaic panel**) is a packaged interconnected assembly of solar cells, also known as photovoltaic cells. The solar panel is used as a component in a larger photovoltaic system to offer electricity for commercial and residential applications.

Because a single solar panel can only produce a limited amount of power, many installations contain several panels. This is known as a photovoltaic array. A photovoltaic installation typically includes an array of solar panels, an inverter, batteries and interconnection wiring.

Solar panels use light energy (photons) from the sun to generate electricity through the photovoltaic effect (this is the photoelectric effect). The structural (load carrying) member of a module can either be the top layer (superstrate) or the back layer (substrate). The majority of modules use wafer-based crystalline silicon cells or a thin-film cell based on cadmium telluride or silicon. Crystalline silicon, which is commonly used in the wafer-form in photovoltaic (PV) modules, is derived from silicon, a commonly used semiconductor.

A group of solar cells joined together in a definite pattern form a panel which can harness a large amount of solar energy and can produce electricity enough to run street light, irrigation water pump, etc.

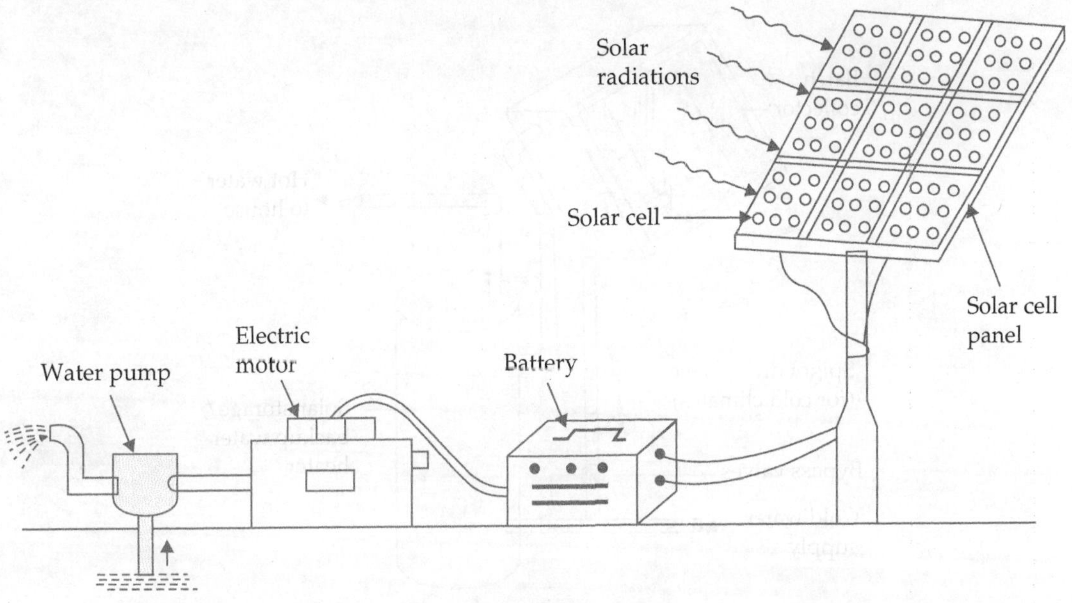

FIGURE 6.16 **Solar panel.**

Applications of solar panels:

In order to use the cells in practical applications, they must be:

- Connected electrically to one another and to the rest of the system.
- Protected from mechanical damage during manufacture, transport, installation and use (in particular against hail impact, wind and snow loads). This is especially important for wafer-based silicon cells which are brittle.
- Protected from moisture, which corrodes metal contacts and interconnects, (and for thin-film cells the transparent conductive oxide layer) thus decreasing performance and lifetime.
- Photovoltaic systems are used for either on- or off-grid applications, and for solar panels on spacecraft.

Solar Water Heaters. Like solar collectors, solar water heaters can be either passive or active. Passive systems have no moving parts, no external energy source except the Sun, and rely on the basic principle of physics that hot water rises and cold water falls. Active systems utilize a mechanical circulating pump and some type of temperature control.

A simple passive solar water heater consists of a water tank that has been painted black and placed in a well-insulated box that has glass or plastic on one side. The sun's rays penetrate the glass and heat the tank. This type of system is often called a "bread box" or batch heater. It allows cold water to flow in at the bottom and hot water to flow out from the top. The system operates on water pressure from the water supply. Water from the system is then transferred to a conventional gas or electric water heater. A thermostat on the water heater determines whether the water is hot enough to be used. If it is not, the conventional water heater adds the required heat.

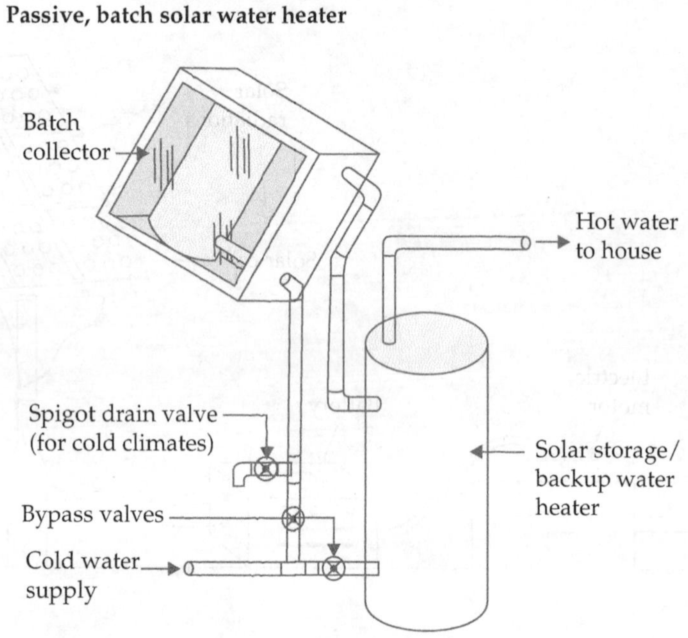

FIGURE 6.17 **Solar water heater.**

Solar Space Heating. A glycol (antifreeze) and water mixture is circulated through the solar collectors to transfer heat, there will be a thermal storage tank that can, based on size, provide enough storage to last overnight (or even for up to or several days) if desired.

As shown above, a fan coil is installed in the existing duct system. Multiple zones can be accommodated. The solar heated glycol/water mix is circulated through the fan coil. There is a low medium speed fan that runs more-or-less continuously but is controlled so that it runs only when the fan coil temperature exceeds indoor temperature, and of course the fan is not allowed run when doing so would overheat the interior space. Another sensor allows the controls to isolate the fan coil when the solar heated water is not hot enough to contribute, so as to keep the system from removing heat from the building.

The result is the delivery of a constant flow of heated air, in the 90–125°F range into the building which reduces or eliminates the need to run the legacy heating source. While this constant flow of heat may at times be not much above human body temperature and therefore not feel "hot" at the vent, it is sufficient to satisfy the thermostat and maintain comfortable conditions within the building. At any time that more heat is needed, the legacy system can provide supplemental heating.

Solar space heating starts with (preferably) southerly facing evacuated tube solar collectors mounted on the roof or other structure. A properly configured system will be able to provide heat to the building with outdoor ambient conditions as low as –23°C.

During the summer, the hot water can be used to provide free hot water to eliminate water heating bills all season long, and/or heat a swimming pool or hot tub. Alternatively, the solar heating system can be disabled.

Parabolic Troughs. A parabolic trough collector has a long parabolic-shaped reflector that focuses the sun's rays on a receiver pipe located at the focus of the parabola. The collector tilts with the sun as the sun moves from east to west during the day to ensure that the sun is continuously focused on the receiver.

Because of its parabolic shape, a trough can focus the sun at 30 to 100 times its normal intensity (concentration ratio) on the receiver pipe located along the focal line of the trough, achieving operating temperatures over 750°F.

Flat Plate Collectors. Flat plate thermal system for water heating deployed on a flat roof.

They consist of a

(*i*) dark flat plate absorber of solar energy.

(*ii*) transparent cover that allows solar energy to pass through but reduces heat losses.

(*iii*) heat-transport fluid (air, antifreeze or water) to remove heat from the absorber.

(*iv*) heat insulating backing.

The absorber consists of a thin absorber sheet often backed by a grid or coil of fluid tubing placed in an insulated casing with a glass or polycarbonate cover. In water heat panels, fluid is usually circulated through tubing to transfer heat from the absorber to an insulated water tank. This may be achieved directly or through a heat exchanger. Most air heat fabricates and some water heat manufacturers have a completely flooded absorber consisting of two sheets of metal which the fluid passes between. Because the heat exchange area is greater they may be marginally more efficient than traditional absorbers.

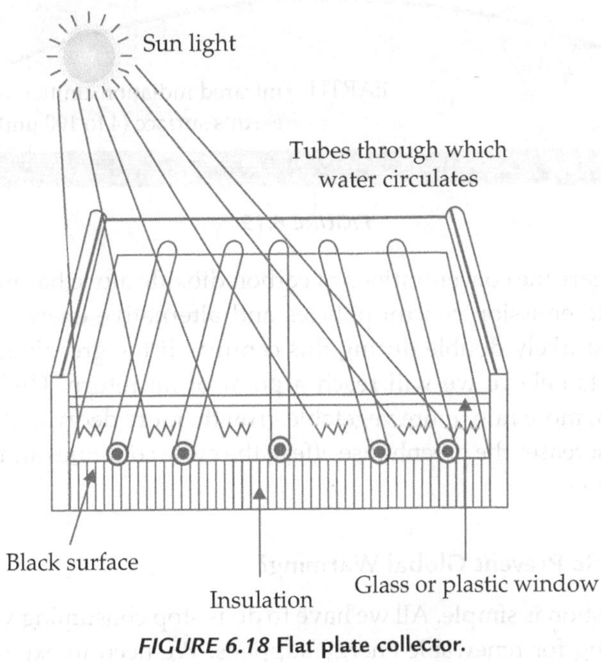

FIGURE 6.18 **Flat plate collector.**

A pool or unglazed collector is a simple form of fat-plate collector without a transparent cover. Typically polypropylene or EPDM rubber or silicon rubber is used as an absorber. Used for pool heating it can work quite well when the desired output temperature is near the ambient temperature (that is, when it is warm outside). As the ambient temperature gets cooler, these collectors become less effective. Most flat plate collectors have a life expectancy of over 25 years.

6.7 GREENHOUSE CONCEPT

Energy from the Sun heats the earth's surface, makes clouds and creates weather. Atmospheric greenhouse gases such as CO_2 and clouds trap some of the outgoing energy like a great blanket. The greenhouse effect moderates temperatures and makes life possible on this small green planet. The threat of ***global warming*** may someday become a serious problem if the current trend of fossil fuel consumption continues.

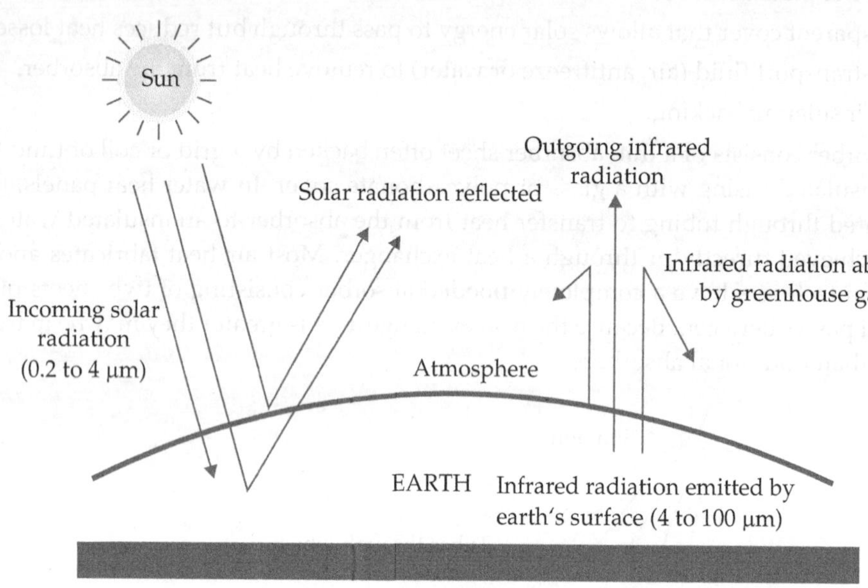

FIGURE 6.19

In the last 300 years the concentrations of carbon dioxide alone has increased by more than 30%. Without adequate emission control policies and alternative energies to choose from this concentration will most likely double during this century. If the greenhouse effect continues to escalate many scientists believe we will reach a point of no return. High temperatures mean increased plant growth, more rain, more vegetable growth, more decay and more carbon dioxide. More carbon dioxide increases the greenhouse effect, the cycle continues and the planet "hot balls" into an inferno like Venus.

6.7.1 How can People Prevent Global Warming?

The answer to this question is simple. All we have to do is stop consuming vast quantities of fossil fuels and start searching for renewable energy supplies. We need to explore alternative energy options such as wind, water and solar energy. Of these options solar energy appears to have the greatest potential over the largest area. To avoid excessive global warming we can use the same principle that's causing it. That's right! The greenhouse effect is a fundamental method of trapping heat from the Sun. High frequency radiation such as ultraviolet and visible light is allowed to pass through a glazing. Once the light is inside the greenhouse and it strikes a non reflective surface the high frequency radiation is transformed into a low frequency heat radiation that cannot pass back out through the glazing. Heat is trapped.

Advantages and Disadvantages:

1. It can work as a backup in case the dams run out of water or are low and if there is a drought.
2. It cost too much money to pay for the plants.
3. The ocean is always there and easy to get to.
4. It reduces oxygen levels in the air and results in an increase of the density of any discharged waste water.
5. Its unlimited source of water.
6. Make more greenhouse gasses so there for more pollution in the atmosphere.
7. It doubles the amount of salt in the ocean's and impacts on the environment for animals and humans.

6.8 GREEN CHEMISTRY

Introduction. To meet the demands of modern civilization, variety of chemical products and industries have evolved which lead to the formation of hazardous substances. In order to prevent or minimize the formation and use of such hazardous substances, the chemists are required to develop the novel technologies which has inspired the generation of new branch of chemistry called green chemistry.

- Basic idea of green chemistry is a new approach which provides the design of chemical products and process that minimize the use and formation of hazardous substances so that it is inherently of less risk to human health and the environment.
- Laboratory chemistry plays a key role to develop green chemical methods to prevent the pollution in the environment. Now, the revolution of green chemistry begins with new challenges for chemists involved with the discovery, manufacture and use of chemicals.

Definition. A branch of chemistry which designs the chemical products and processes that minimize the use and formation of hazardous substances is known as green chemistry.

6.8.1 The Need of Green Chemistry

The 20th century is very successful and highest scientific development with respect to various benefits to the mankind, but all this was responsible for a number of environmental problems at the local and global levels.

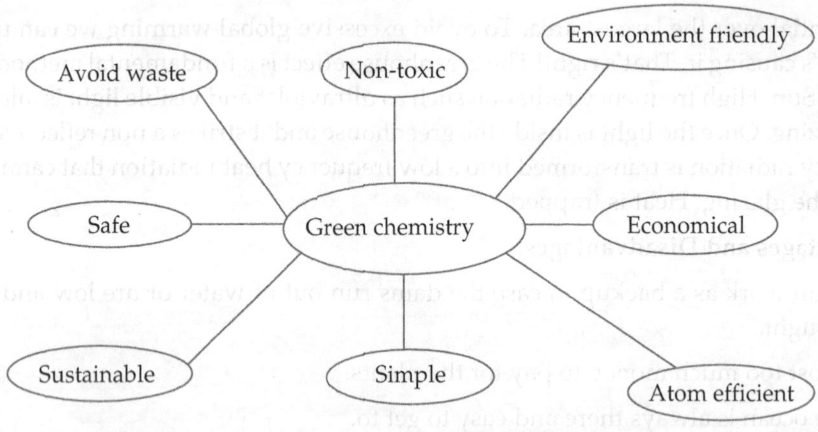

FIGURE 6.20

- Industrial discharges, pesticides and herbicides fill the skies and seas.
- Our technological power is increased day by day but the side effects and potential hazardous also increased.
- To maintain our current standard of living and improve our quality of life, society has come to depend on the products of chemical industry.
- Now a day the public is equally aware of hazardous substances used and generated by the chemical processes.
- Our environment, which is endowed by nature, needs to be protected from ever increasing chemical pollution associated with contemporary life styles and emerging technologies.
- Developments in water treatment, waste disposal method agricultural pesticides and fungicides, polymers, material sciences, detergents, petroleum, additives and so forth have all contributed to the improvement in our quality of life.
- Green chemistry concept is to avoid toxic substance, reduce energy requirement, and use renewable materials.
- Green chemistry includes any chemical process or technology that improves the environment and thus our quality of life.
- The risk of human health and environment can be eliminated through fundamental chemical methodologies in green chemistry.

6.8.2 The Principles of Green Chemistry

Chemistry plays an important role in our lives like clothes, food, building, etc., on the Earth. But in another way it leads to the adverse effects of land, water and atmosphere. This is caused mainly due to effects of byproducts such as pollution of land, water and atmosphere. One of the most important new aspects of chemistry is the development of green chemistry. Green chemistry has emerged as an important aspect of all chemistry. Green chemistry is the design of chemical products and processes that reduce or eliminate the use and generation of hazardous substances.

Green Chemistry is Based on Twelve Principles:

1. Minimization of the waste product formation or prevention. The first principle aims to develop zero waste technology (ZWT). In terms of ZWT, in a chemical synthesis, waste product should be zero or minimum. It also aims to use the waste product of one system as the raw material for other systems. As for example, bottom ash of thermal power station can be used as a raw material for cement and brick industry.

2. Atom economy. Synthetic methods should be designed to maximize the incorporation of all materials used in the process into the final product or to minimize the formation of by product.

$$\text{Percentage atom economy} = \frac{\text{Mass of desired product}}{\text{Total mass of reactants}} \times 100\%$$

3. Less hazardous chemical syntheses. Wherever practicable, synthetic methods should be designed to use and generate substances that possess little or no toxicity to human health and the environment.

4. Designing safer chemicals. Chemical products should be designed to effect their desired function while minimizing their toxicity.

5. Safer solvents and auxiliaries. The use of auxiliary substances (*e.g.*, solvents, separation agents, etc.) should be made unnecessary wherever possible and innocuous when used.

6. Design for energy efficiency. Energy requirements of chemical processes should be recognized for their environmental and economic impacts and should be minimized. If possible, synthetic methods should be conducted at ambient temperature and pressure.

7. Use of renewable feedstock's. A raw material or feedstock should be renewable rather than depleting whenever technically and economically practicable.

8. Reduce derivatives. Unnecessary derivatization (use of blocking groups, protection/deprotection, temporary modification of physical/chemical processes) should be minimized or avoided if possible, because such steps require additional reagents and can generate waste.

9. Catalysis. Catalytic reagents (as selective as possible) are superior to stoichiometric reagents.

10. Design for degradation. Chemical products should be designed so that at the end of their function they break down into innocuous degradation products and do not persist in the environment.

11. Real-time analysis for pollution prevention. Analytical methodologies need to be further developed to allow for real-time, in-process monitoring and control prior to the formation of hazardous substances.

12. Inherently safer chemistry for accident prevention. Substances and the form of a substance used in a chemical process should be chosen to minimize the potential for chemical accidents, including releases, explosions, and fires.

6.8.3 Methods for Green Synthesis or Green Reactions

Chemistry plays an important role to develop our quality life and achieving sustainability on earth. All over the world using advanced knowledge to develop new synthetic conditions like aqueous phase, solvent free, microwave induced, etc. These are produced pollution free environment.

Following are given some of the methods for greener synthesis:

1. Aqueous Phase Method for Green Synthesis. Environmental pollution is mainly caused by organic solvents. So now the all over world have been trying to carry out organic reactions in aqueous phase. The advantages of using water as a solvent is its low cost, non-flammable, devoid of any carcinogenic effects, high specific heat resistance, unique enthalpic and entropic properties and easily handling.

Example. Knoevenagel reactions:

- The condensation of carbonyl compounds with active methylene compounds in presence of weak base like ammonia/amine/pyridine is known as Knoevenagel reaction.

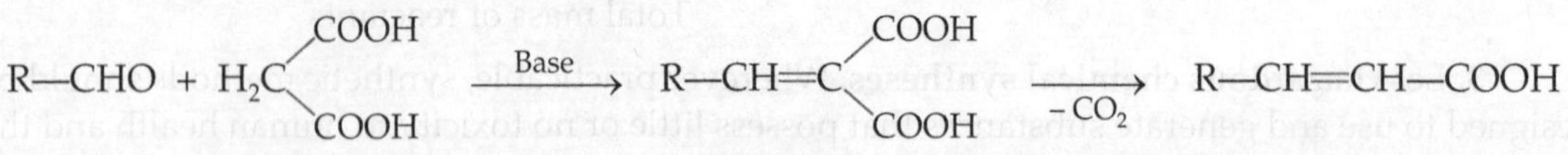

Aldehyde Melonic acid

- The Knoevenagel reaction has been carried out in water medium between aromatic aldehydes and acetonitrile.

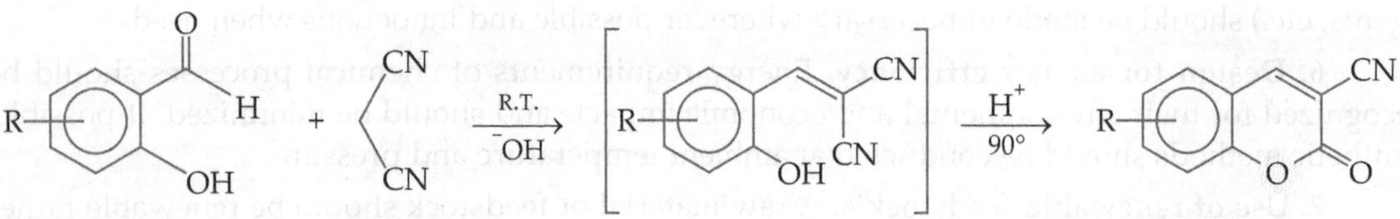

Salicyaldehydes R=H, OH, OMe Malono nitrile 3-cyano cumarins 75–95%

2. Supercritical Fluid Extraction Method for Green Process. Organic solvents have been used in conventional solvent extraction process for extraction of foods, caffeine, and cholesterol and in number of processes like petrochemical processes polymer processes, etc. But these solvent extraction processes are become cause for pollution of land, air and water. To avoid these pollution problems a new technology was emerged in mid of 20th century with large number of potential applications in industry for extraction processes with concern of pollution free environment that is supercritical fluid extraction process.

- Supercritical fluid is termed as it is neither gas nor a liquid but it has the properties of gases and liquids.
- These are produced by heating of gas above its critical temperature or compressing a liquid above its critical pressure.

Example. CO_2 . C_2H_2, NH_3, H_2O, cyclohexane, *n*-pentene, etc.

Principle. The basic principle of SCF extract is that when the feed material is contacted with a supercritical fluid then the volatile substances will partition into the supercritical phase. After dissolution of soluble material the SCF containing the dissolved substances is removed from the feed material. After the separation of extract from SCF may then be recompressed to the extraction conditions and recycled.

Advantages:

1. Dissolving power of the SCF is controlled by pressure or temperature.
2. SCF is easily recoverable from the extract due to its volatility.

3. High flexibility of processing conditions.
4. Easy optimization of the quality and yield of the desired product.
5. Removal or undesirable components and effective concentration of desirable components.

Applications:

1. Used for extraction of essential oils, flavours and some natural products from natural plants.
2. Used in petrochemical industries for distillation residue of the crude oil.
3. Used in extraction of by products from polymer components.
4. Used in purification of surfactants.
5. Used in removal of caffeine from coffee.

3. Phase transfer catalyst (PTC). Phase transfer catalyst is heterogeneous catalyst, which is used to solubilise salts which are insoluble in organic phase solvent. The PTC catalyst can facilitate the migration of the reactants from one phase into another phase where reaction occurs.

- By using PTC, one can achieve faster reactions obtained higher yield by convection of products, make less by products to eliminate the need of expensive and dangerous solvents. So, PTC is especially need for green chemistry.

For example, quaternary ammonium salts like benzyl tri methyl ammonium chloride, phosphonium salts like Hexa decyl tri butyl phosphonium bromide and crown ethers.

Involvement of PTC in reactions given below,

$$\underset{\text{Organic phase}}{R\text{–}Y} + \underset{\text{Aqueous phase}}{X^-} \xrightarrow[\text{PTC}]{Q^+} R\text{–}X + Y^-$$

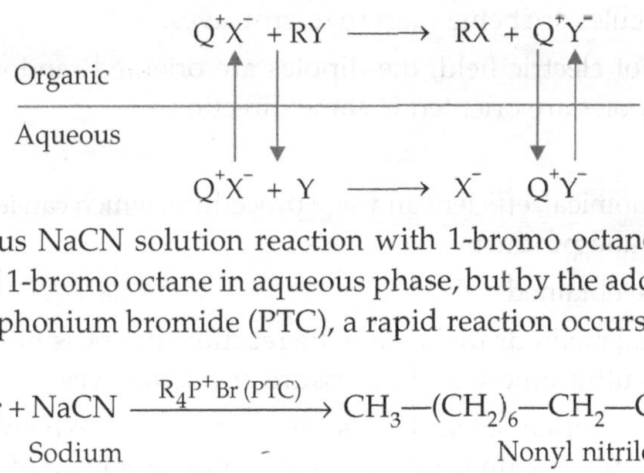

Example. Aqueous NaCN solution reaction with 1-bromo octane does not readily occur due to poor solubility of 1-bromo octane in aqueous phase, but by the addition of small amount of hexadecyl tributyl phosphonium bromide (PTC), a rapid reaction occurs to give nonyl nitrile.

$$\underset{\text{1-bromo octane}}{CH_3\text{—}(CH_2)_6\text{—}CH_2\text{—}Br} + \underset{\text{Sodium cyanide}}{NaCN} \xrightarrow{R_4P^+Br\ (PTC)} \underset{\text{Nonyl nitrile}}{CH_3\text{—}(CH_2)_6\text{—}CH_2\text{—}CN} + NaBr + R_4P\,Br(PTC)$$

4. Biocatalyst for Green Synthesis. Biocatalysis can be defined as a traditional chemical catalysis and natural substances, which can be one or more enzymes or cells, living, catalyze a chemical reaction or series of chemical reactions.

- Biocatalysis includes one step enzymatic conversion like fumare to aspartic acid, two-step conversions like ethanol to acetic acid and multi-step conversions like molasses to ethanol.
- It is well known that most of the antibiotics have been prepared using biocatalysts.

(*i*) **Oxido reductase.** These enzymes catalyze oxidation and reduction reactions.

(*ii*) **Transferases.** These enzymes catalyze the transfer of various functional groups.

(*iii*) **Lyases.** These aid two types, one which catalyse addition to double bond and other which catalyses removal of groups and leaves double bond. Both addition and elimination of small molecules are on sp^3 carbon.

(*iv*) **Isomerases.** These catalyze various types of isomerization.

Example:

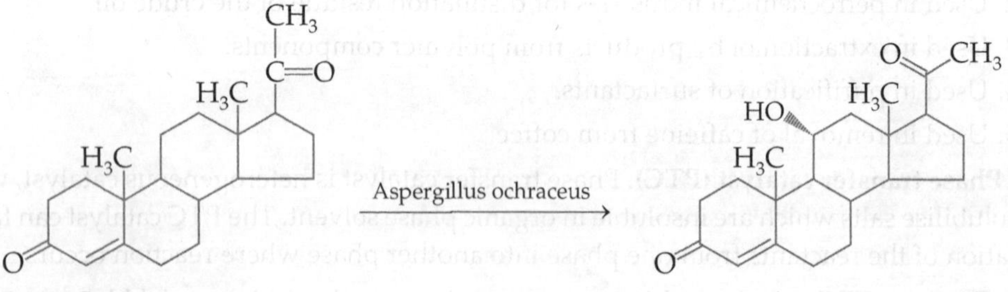

Progesterone 11 α-hydroxy progesterone

5. Microwave Induced Method for Green Synthesis. Microwaves have wavelength range from 1 cm to 1 m in the electromagnetic spectrum and are located between the infrared and radio/radar-frequencies.

- The main purpose of microwaves is used for heating.
- According to Plancks law ($E = h\nu$), the quanta energy involved can be evaluated to 0.3 cal/mole. This energy is too low to induce a molecule for extraction of any reaction. The main source to exposure ahead under microwaves is microwave interactions.
- Polar molecules can involve in selective absorption of electromagnetic microwaves but non polar molecules are being inert to microwaves.
- In the absence of electric field, the dipoles are oriented randomly; in the presence of electric field dipoles are oriented in same direction.

Advantages:

1. It is a clean, economical, efficient and safe procedure, which can lead to substantial saving in money, time and products.
2. Better yields are obtained.

Unlike conventional methods for heating the reaction, it is possible to activate processes by physical methods such as ultrasonic sounds, pressure or microwaves.

The reactions can be carried out under microwaves method. Whether in solvent medium or non solvent medium. Limited reactions in solvent medium, and most of the reactions in solvent free medium.

There are three different types in using microwave reactions.

(*i*) **Microwave solvent–free reactions (solid state reactions).** Solvent free technique hold a strategic position as a solvent is very often toxic, expressive and problematic to use and to remove. It is the main reason for development of such modern technologies. These approaches can also enable experiments avoiding the use of strong safety and pollution problems.

These acids can be replaced by solid recyclable acids such as clays. These types of reactions are known as solid state reactions.

Example:

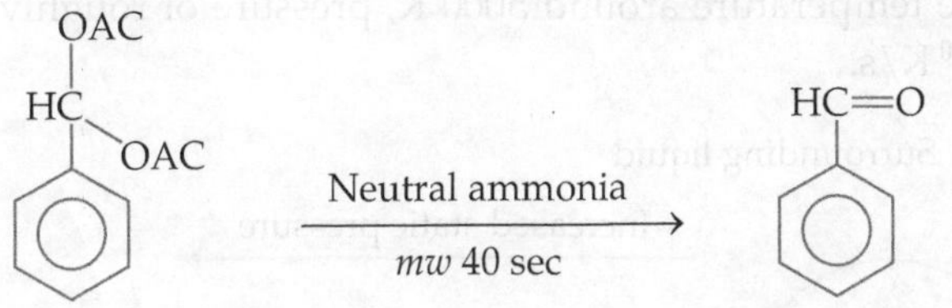

(*ii*) **Microwave-assisted reaction in organic solvents.** Microwaves have been used for synthesis of chalcones and related enones in presence of organic solvents.

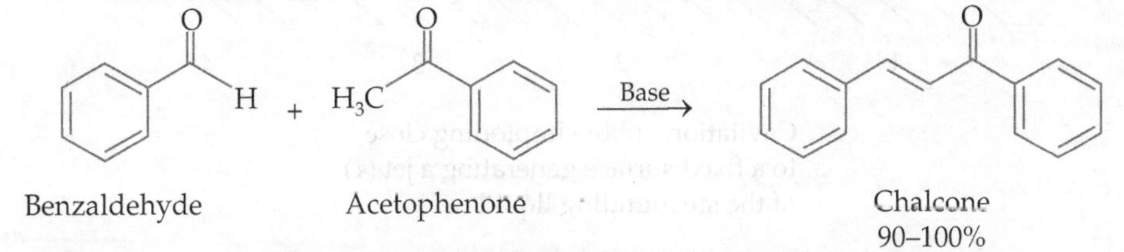

(*iii*) **Microwave assisted reactions in aqueous phase Hoffman elimination reactions.** In normal method quaternary ammonium salts are heated at high temperature and the yield of the products is low. Use of microwave irradiation has led to high yielding synthesis of a thermal sensitive Hoffman elimination product in water-chloroform system.

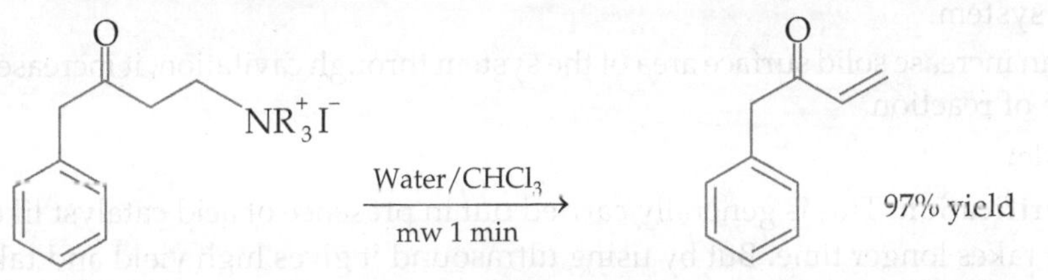

6. Ultrasound Assisted Method for Green Synthesis. Sound waves having high frequencies are known as ultrasound waves. (about 20–100 kHz). The ultrasound is generated with an instrument having an ultrasonic transducer, a device by which electrical or mechanical energy can be converted into sound energy.

Commonly used ultrasonic transducer is "quartz" but in modern ultrasonic equipment, ceramic impregnated barium titanate used as transducer which converts. Over 95% of electrical energy into ultrasound.

These are worked based on the "piezoelectric effect" it is nothing but a equal and reversing charges are applied to opposite faces of the transducer resulted generation of vibrations by expansion and contraction of transducer which emits ultrasonic waves.

The term *sonochemistry* is used to describe the effect of ultrasound waves on chemical reactivity. This depends upon phenomenon of "sonic cavitations or acoustic cavitations".

Upon irradiation with high intensity sound or ultrasound waves, which are propagated a series of compressing refraction cycles; pass through a liquid medium results oscillation of molecules takes place. Under appropriate condition, the attractive forces of the molecules may be overcome, causing formation of bubbles, growth and collapsed which is due to great internal forces in the bubbles. Bubble collapse in liquid produces enormous amount of energy from the conversion of

K.E. of the liquid motion into heating the contents of the bubble. Experimental results have shown that these bubbles have temperature around 5000 K, pressure of roughly 1000 atm and heating, cooling rates above 10^{10} K/s.

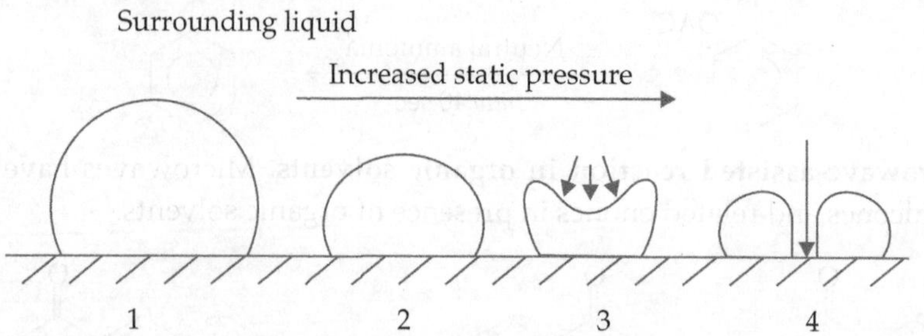

Cavitation bubble imploding close to a fixed surface generating a jet (4) of the surrounding liquid

FIGURE 6.21

Advantages:

1. It can enhance chemical reactivity.
2. It can effectively activate the catalyst by excitation of the atomic and molecular modes of the system.
3. It can increase solid surface area of the system through cavitation, it increases the observed rate of reaction.

Example:

***(i)* Esterification.** This is generally carried out in presence of acid catalyst like H_2SO_4, gives low yield and takes longer time. But by using ultrasound it gives high yield and takes short time.

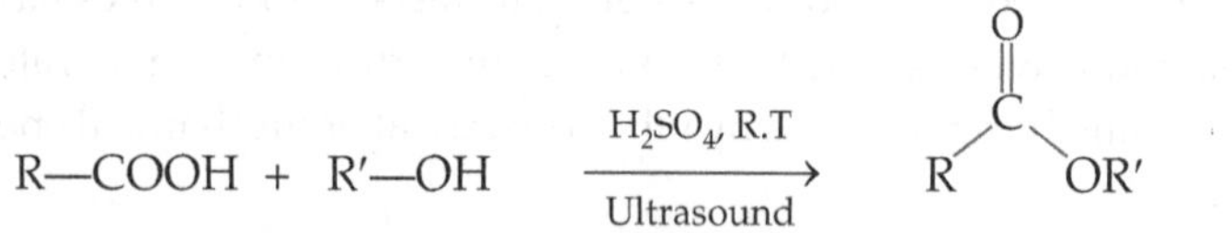

***(ii)* Alkylation.** N-alkylation of Indole takes place under sonication in the presence of a PTC reagent polyethylene glycol and monoethyl ether with better yield.

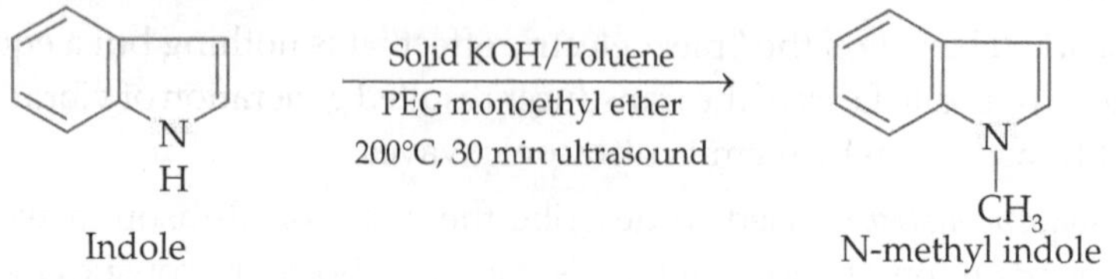

Indole N-methyl indole

7. Ionic Liquid as Green Solvent Method. The commonly used solvents for organic synthesis are DMF, Benzene, Toluene, CFC, methylene chloride, etc. It is well known that these are cause for health and environmental problems such as ozone depleting, greenhouse effect, carcinogenic, toxic and decomposition of these solvents takes place at higher temperature leads to noxious compounds.

- Usually ionic liquids are organic salts, which often exhibit physical properties close to fluids or liquids at near ambient temperatures.

Example the first example of ionic liquid is ethyl ammonium nitrate (m.p. 120°C). Today, there are several ionic liquids with different combinations of cations and anions are prepared.

Example:

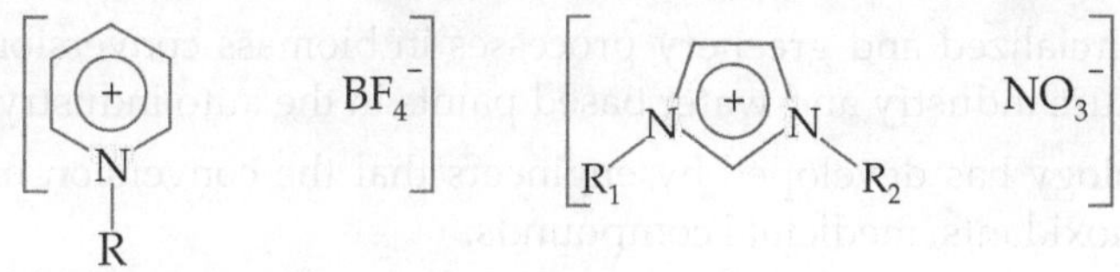

Advantages and properties of ionic liquids:

1. These are having negligible vapour pressure.
2. Ionic liquid have tunable physical chemical and electrical properties.
3. They are non flammable.
4. They can be relatively in expensive to synthesis used as solvents and catalysts.

Example:

Synthesis of pharmaceutical compounds. Ionic liquids have been used in the synthesis of pharmaceutical compounds. A representative example is the synthesis of pravadoline.The method consists the alkylation of 2-methyl indole with 1-(N-morpholino), 2-chloroethane in PF_6 to give 95% yield of the corresponding N-alkyl derivative. Subsequent Friedel craft reaction with *p*-methoxy benzoyl chloride in chloro aluminium(III) ionic liquid gives pravadoline.

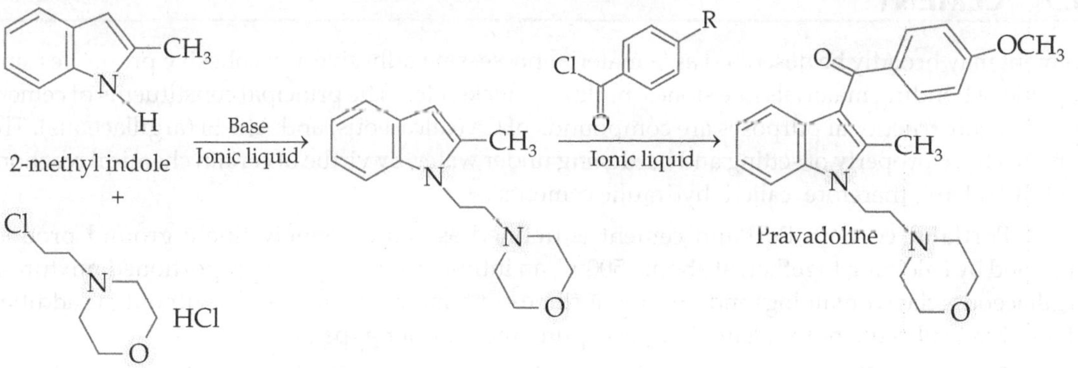

1-(N-morpholino), 2-chloroethane

6.8.4 Protection of the Environment by Science and Technology

Or

Engineering Applications of Green Chemistry

1. Application of ecofriendly engineering principles to the design of manufacturing and combusting processes to reduce air pollutant emissions to acceptable levels.
2. Developing of various processes to remove nitrogen oxides, sulphur oxides, volatile organic compounds (VOC), reactive organic gases and other air pollutants form fuel gases.
3. Air pollution dispersion modes have to develop to remove air pollutants like vehicle exhausting gases, industrial flue stack gases.

4. Improving industrial processes to eliminate waste and reduce consumption of organic solvents.
5. Engineers have been trying to develop natural resources like forest, lakes, streams and wetlands, green buildings and cleaning up polluted waste sites.
6. A new commercialized and greenery processes in biomass conversion used in synthetic fabric in furniture industry and water based paints in the auto industry.
7. New methodology has developed by engineers that the conversion of lignin into useful chemicals, antioxidants, medicinal compounds.
8. The main aim of the nanotechnology is to minimize the environmental and human health risk associated with the exits ion manufacture products.
9. By using biotechnology processes. The biomasses converted to fermentable sugars in presence of enzymes from crop residues such as corn Stover and stalks, wheat straw and rice straw.
10. Bio polymers developed by engineers like solanyle which is produced from byproduct of the potent starch industry.
11. Engineers established chemical production processes, which have led to reduce energy, water consumption, minimized by products and even noise reduction.
12. The engineers provided chemical processes, which give adverse array of valuable products and materials for health care.

6.9 CEMENT

Cement may broadly be described as 'a material possessing adhesive and cohesive properties, and capable of bonding materials like stones, building blocks, etc. "The principal constituents of cement used for constructional purposes are compounds of Ca (calcareous) and Al + Si (argillaceous). The cements have property of setting and hardening under water, by virtue of certain chemical reactions with it and are, therefore, called 'hydraulic cements'.

Portland cement. Portland cement is defined as "an extremely finely ground product obtained by calcining together, at about 1500°C, an intimate and properly proportioned mixture of argillaceous (clay-containing) and calcareous (lime-containing) raw materials, without the addition of anything subsequent to calcinations, excepting the retarder gypsum.

Properties. Portland cement is the most important and reliable cementing material, used for constructional works. It was discovered in 1824 by Joseph Aspidin, a brick layer of Leeds. It was so-named, because a part of cement with water on setting and hardening resembled in colour and hardness to portland stone, a limestone quarried in Dorset.

6.9.1 Manufacture of Portland Cement

Raw materials. Raw materials for the manufacture of cement are:

1. Calcareous materials, CaO (such as limestone, chalk, marl, etc.),
2. Argillaceous materials, Al_2O_3 and SiO_2 (such as clay, shale, slate, etc.),
3. Powdered coal or fuel oil, and
4. Gypsum ($CaSO_4 \cdot 2H_2O$).

Functions of the ingredients of cement:

1. Lime is the principal constituent of cement. Its proportion must be properly regulated. However, excess of lime reduces the strength of cement, because it makes the cement to expand and disintegrated. On the other hand, presence of lesser amount of lime than required also reduces the strength of cement and makes it quick-setting.
2. Silica imparts strength to cement.
3. Alumina makes the cement quick-setting. Excess of alumina, however, weakens the cement.
4. Calcium sulphate (gypsum) helps to retard the setting action of cement. It actually enhances the initial setting time of cement.
5. Iron oxide provides: (*a*) colour, (*b*) strength, (*c*) hardness to the cement.
6. Iron oxide provides, in small proportion, is desirable. When present in small amount, it imparts soundness to cement. However, its excess reduces the soundness of cement.
7. Alkalis, if present in excess, cause the cement effervescent.

Manufacturing of Portland Cement Involves the Following Steps:

1. Mixing of raw materials can be done either by: (*a*) dry process or (*b*) wet process.

(*a*) **Dry process.** The raw materials limestone (or chalk) and clay (or shale) are crushed (in gyratory crushers) into roughly 2–5 cm size pieces. Then, these are ground to fine powder (in ball mills/tube mills). Each separate powdered ingredient is stored in a separate hopper. Then, the powdered materials are mixed in the requited proportions to get dry 'raw mix', which is stored in storage bins (called silos) and kept ready to be fed in a rotary kiln. Raw materials are mixed in calculated proportions so that the average composition of the final product is as follows:

Component	*Percentage of mass*
Lime (CaO)	60–69
Silica (SiO_2)	17–25
Alumina (Al_2O_3)	3–8
Iron oxide (Fe_2O_3)	2–4
Magnesium oxide (MgO)	1–5
Sulphur trioxide (SO_3)	1–3
Alkali oxides ($Na_2O + K_2O$)	0.3–1.5

(*b*) **Wet process.** The calcareous raw materials are crushed, powdered and stored in big storage tanks (called silos). The argillaceous material (say clay) is thoroughly mixed with water in wash mills, to remove any adhering organic matter, etc. The basin-washed clay is also stored. Powdered limestone (from silos) and washed wet-clay (from basins) are allowed to flow in a channel, in the right proportions. From the channel, the two raw materials are led to 'grinding mills', where they are mixed intimately to form a paste, called slurry [Grinding operation is carried out in either tube mill or ball mill or both]. The slurry is led to a 'correcting basin', where its chemical composition may be adjusted, if necessary. This slurry contains about 38 to 40 per cent water. The slurry is finally stored in storage tanks and kept ready for feeding to a rotary kiln.

Merits and demerits of dry and wet process:

Dry process	Wet process
1. It is adopted when the raw materials are quite hard.	It can be used for any type of raw materials.
2. Fuel consumption is low.	Fuel consumption is higher.
3. Process is slow.	Process is comparative faster.
4. Cement produced is of inferior quality	Cement produced is of superior quality.
5. Cost of production of cement is less.	Cost of production of cement is somewhat higher.
6. On the whole, the process is costly.	On the whole, the process is cheaper.

2. Burning. Burning is usually done in rotary kiln, which is a steel tube, about 2.5 to 3.0 m in diameter and 90 to 120 m in length, lined inside with refractory bricks. The kiln is laid in slightly inclined position at a gradient of 1 in 25 to 1 in 30. The rests on roller bearings, which are supported on column of masonry or concrete. The kiln is capable of rotating at 1 r.p.m. (revolution per minute) about its longitudinal axis. Burning fuel (usually powdered coal or oil) and air are injected at the lower end. A long hot flame is produced, which heats the interior of the kiln up to a maximum temperature of about 1,750°C.

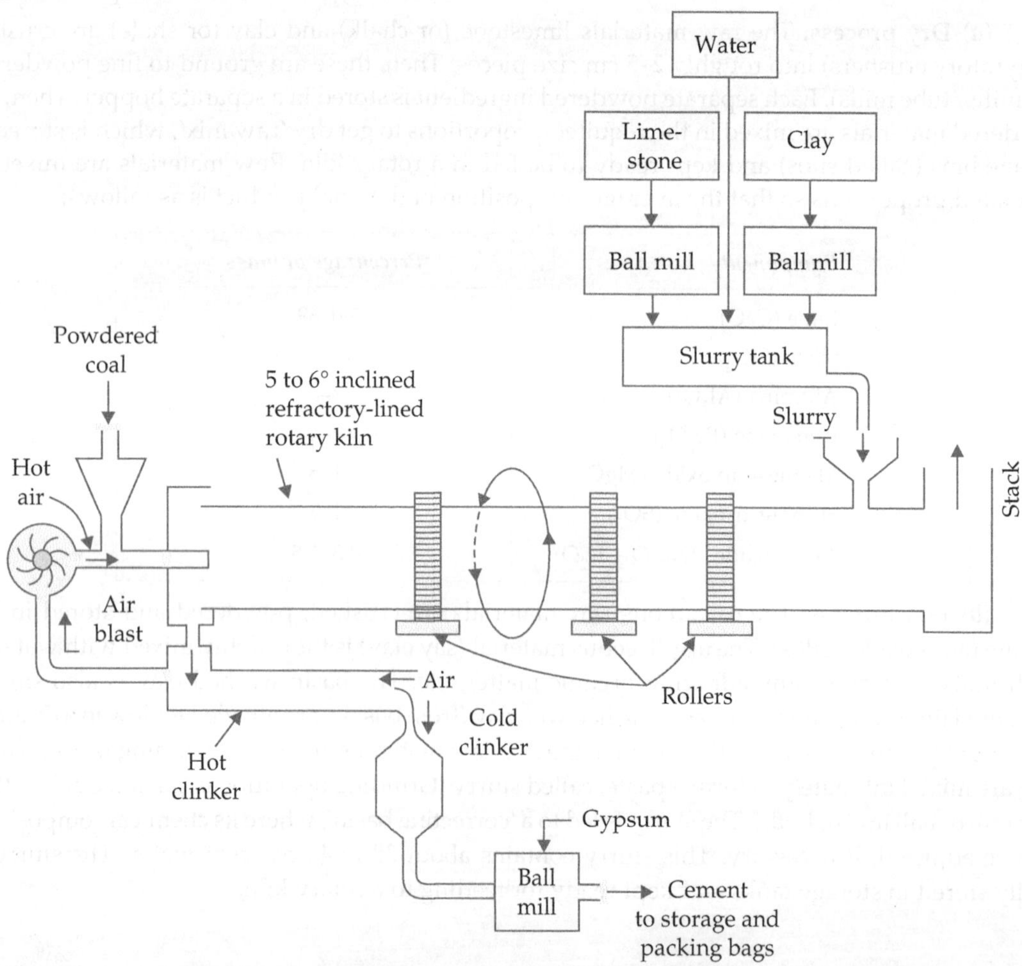

FIGURE 6.22 **Rotary cement kiln.**

Process. The 'raw-mix' or 'corrected-slurry' is injected into the kiln at its upper end; while hot flames is forced into the kiln from the lower end. Due to slope and slow rotation of the kiln, the materials fed in move continuously towards the hottest-end at a speed of about 15 m per hour. As the mixture or slurry gradually descends, the temperature rises.

Chemistry:

(*i*) In the upper part of kiln, where the temperature is around 400°C, most of the water in the slurry gets evaporated (drying zone).

(*ii*) In the central part of kiln, where the temperature is around 1,000°C, limestone of dry mix or slurry undergoes decomposition to form quicklime and carbon dioxide; and the latter escapes out. The material forms small lumps, called nodules [calcinations zone].

$$\underset{\text{Limestone}}{CaCO_3} \rightleftharpoons \underset{\text{Quicklime}}{CaO} + CO_2\uparrow$$

(*iii*) In the lower part of the kiln, the temperature is between 1,500 to 1,700°C. Here lime and clay (of nodules) undergo chemical interaction or fusion, yielding calcium aluminates and silicates (clinkering zone).

$$2CaO + SiO_2 \longrightarrow \underset{\text{Dicalcium silicate } (C_2S)}{Ca_2SiO_4}$$

$$3CaO + SiO_2 \longrightarrow \underset{\text{Tricalcium silicate } (C_3S)}{Ca_3SiO_5}$$

$$3CaO + Al_2O_3 \longrightarrow \underset{\text{Tricalcium aluminate } (C_3A)}{Ca_3Al_2O_6}$$

$$4CaO + Al_2O_3 + Fe_2O_3 \longrightarrow \underset{\text{Tetracalcium aluminoferrite } (C_4AF)}{Ca_4Al_2Fe_2O_{10}}$$

The aluminates and silicates of calcium then fuse together to form small (of about 0.5 to 1 cm) diameter hard, grayish stones, called clinkers. These clinkers are very hot (at about 1,000°C). The rotary kiln at the base is provided with another small rotary kiln. In this, hot clinkers fall and cool air is admitted from opposite direction. Air counter-blast cools the clinkers. Hot air s-produced is used for burning powdered coal/oil. The cooled clinkers are collected in small trolleys.

3. Grinding. The cooled clinkers are ground to fine powder in ball mills or tube mills. During final grindings, a small quantity (2–3%) of powdered gypsum is added, so that the resulting cement does not set very quickly, when it comes in contract with water. Gypsum, thus, acts as a retarding agent for early setting of cement.

After the initial set, the cement-water paste becomes stiff, but gypsum retards the dissolution of C_3A by forming tricalcium sulphoaluminate, ($3CaO \cdot Al_2O_3 \cdot xCaSO_4 \cdot 2H_2O$), which is insoluble. Thus,

$$\underset{\text{After initial set}}{3CaO + Al_2O_3} + \underset{\text{Gypsum}}{xCaSO_4 \cdot 2H_2O} \longrightarrow \underset{\substack{\text{Tricalcium sulphoaluminate} \\ \text{(Insoluble)}}}{3CaO \cdot Al_2O_3 \cdot xCaSO_4 \cdot 2H_2O}$$

Formation of insoluble tricalciumsulphoaluminate prevents too early further reactions of setting and hardening.

4. Packing. The ground cement is stored in silos, from which it is fed to automatic packing machines. Each bag, usually, contains 50 kg of cement.

Flow diagram for the manufacturing of portland cement

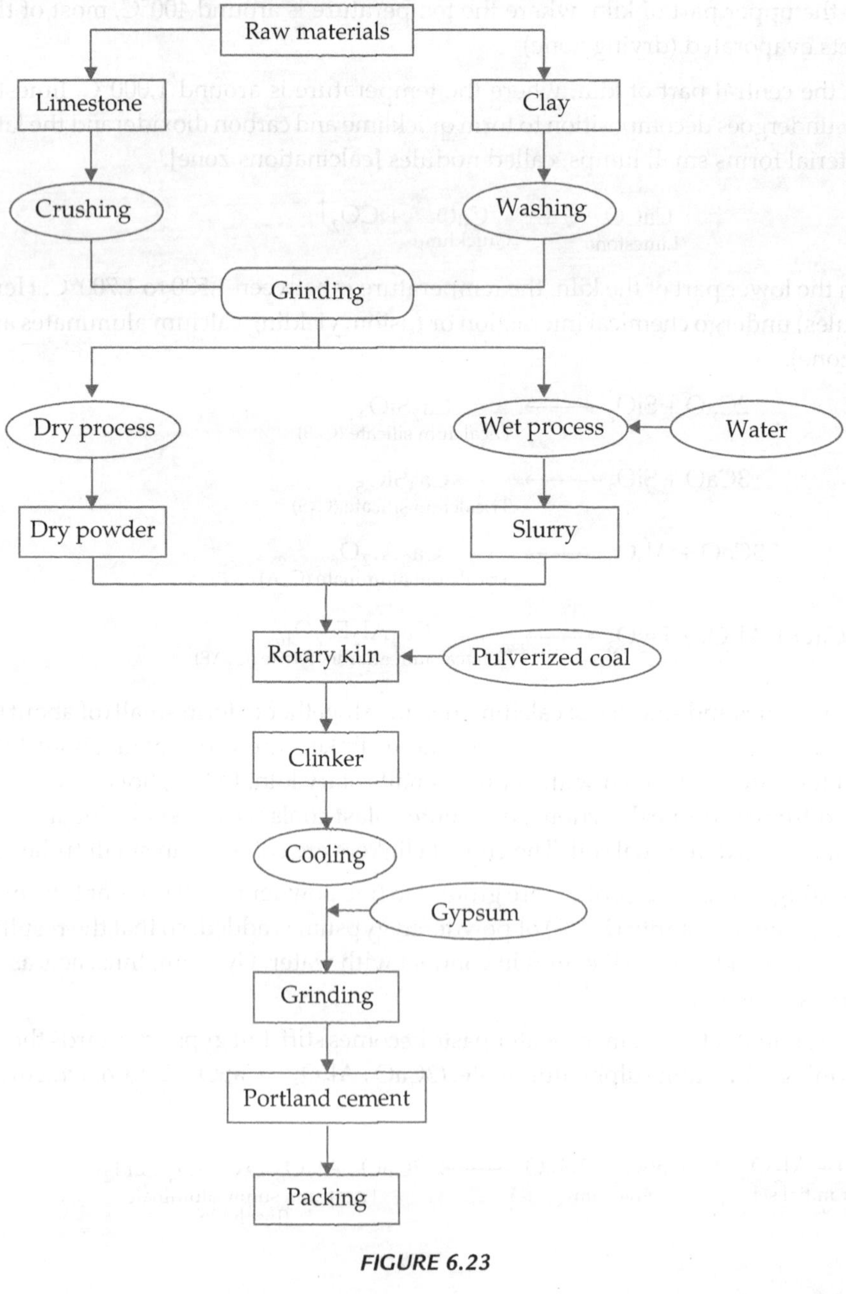

FIGURE 6.23

6.9.2 Chemistry of Setting and Hardening of Portland Cement

When cement is mixed with water to plastic mass, called "cement paste", hydration reaction begins, resulting in the formation of gel and crystalline products. The interlocking of the crystals, finally bind the inert particles of the aggregates into a compact rock-like material. The process of solidification-comprises of: (*i*) setting, and (*ii*) hardening. 'Setting' is defined as stiffening of the original plastic mass, due to initial gel formation. Hardening is development of strength, due to crystallization.

After setting, hardening starts, due to the gradual progress of crystallization in the interior of the mass. The strength developed by cement paste at any time, depends upon the amount of gel formed and the extent of crystallization.

Initial setting. Initial setting of cement-paste is mainly due to the hydration of tricalcium aluminate (C_3A) and gel formation of tetracalcium aluminoferric.

$$C_4AF + 7H_2O \longrightarrow C_3A\,.\,6H_2O\ (\text{crystalline}) + CF.\ H_2O(gel) - 420\ \text{kJ/kg}$$

$$C_3A - 6H_2O \longrightarrow C_3A\,.\,6H_2O + 880\ \text{kJ/kg}$$

$$2C_2S + 4H_2O \longrightarrow C_3S_2\,.\,6H_2O + Ca(OH)_2 - 250\ \text{kJ/kg}$$

Tobermonite gel

Final setting. Final setting and hardening of cement-paste is due to the formation of tobermonite gel plus crystallization of calcium hydroxide and hydrated tricalcium aluminate.

$$2[3CaO\,.\,SiO_2] + 6H_2O \longrightarrow \underset{\text{Tobermonite gel}}{3CaO\,.\,2SiO_2\,.\,3H_2O} + 3Ca(OH)_2 + 500\ \text{kJ}$$

Sequence of chemical reactions during setting and hardening of cement. When water is added to cement, its various constituents undergo hydration and crystallization at different rates. (*i*) At first, hydration of tricalcium aluminate (C_3A) and tetracalcium aluminoferrite (C_4AF) takes place. (*ii*) Next, the hydration of tricalcium silicate (C_3S) begins within 24 hours and gets completed in 7 days. (*iii*) The gel of aluminate begins to crystallize and at the same time, dicalcium silicate (C_2S) begins to hydrate in 7 to 28 days. Thus, the initial set of cement is due to the hydration of aluminate. The development of carly-strength, between 1 to 7 days, is due to the hydration of tricalcium silicate and the further hydration of aluminate. The increase of strength, between 7 to 28 days, is due to hydration of dicalcium silicate and continued hydration of tricalcium silicate.

Function of gypsum in cement. Tricalcium aluminate (C_3A) combines with water very rapidly with the evolution of a large amount of heat.

$$C_3A + 6H_2O \longrightarrow C_3A\,.\,6H_2O + \text{Heat}$$

After the initial-set, the paste becomes somewhat stiff. However, the added gypsum retards the dissolution of C_3A by forming insoluble calcium sulphoaluminate.

This reaction prevents high concentrations of alumina in the cement solution, thereby retarding the early initial-set of the cement.

6.9.3 Heat of Hydration of Cement

When water is mixed with cement, hydration, hydrolysis and gelation reactions start and some heat is liberate simultaneously. On an average, 500 kJ/kg of heat is evolved during complete hydration of cement. The heats of hydration of various constituents of cement are as follows:

Constituent of cement	C_3A	C_3S	C_4AF	C_2S
Heat of hydration (kJ/kg)	880	500	420	250

Consequently, in order to avoid the shrinkage cracks on setting and hardening, it is quite necessary to pour concrete in position in such amounts that the heat liberated during hydration is dissipated as quickly as possible.

6.9.4 Decay of Cement (Effect of CO_2, SO_2 and Chloride on Concrete)

Effect of CO_2. The carbon dioxide (CO_2) is a gas in the atmosphere. It can be dissolved by the concrete pore solution, and react with some calcium compounds to form carbonates (carbonation). So, the pH is about 9 for the pore solution of a concrete deteriorated by carbonation.

The penetration of carbon dioxide in concrete is a diffusion process. It is fast when the concrete is rather dry. But the carbonation reaction does not occur if no pore solution is present in concrete. This is the reason why, the most favourable conditions to the penetration of carbon dioxide correspond to a relative humidity of about 65%.

Effect of SO_2. This is the more common type and typically occurs where water containing dissolved sulfate penetrates the concrete. A fairly well-defined reaction front the composition and microstructure of the concrete will have changed.

- Extensive cracking.
- Expansion.
- Loss of bond between the cement paste and aggregate.

Effect of Chloride. Chlorides are highly soluble in water. So, these ions are dissolved in water, and penetrate with it in concrete (chlorides penetration), either by the wetting of a non-saturated concrete (convection), or by diffusion because the chloride content is higher in the environment than in the original concrete (concentration gradient). The main part of chlorides coming from outside remain as dissolved ions, in the concrete pore solution. But they can also react with some components of the material (chemical reaction or adsorption).

EXERCISE

1. What are carbon nanotubes (CNTs)? Write about SWNTs and MWNTs with structure.
2. How the CNTs are prepared by the following methods?
   (*a*) Arc-discharge method
   (*b*) Laser-ablation method
   (*c*) CVD method.
3. What are nanomaterials? Write about the properties and applications of nanomaterials.
4. Write a note on fullerenes.
5. What are liquid crystals? Explain about the types of liquid crystals.
6. Write the applications of liquid crystals in LCD and engineering.
7. Write a short note on FRPs.

8. Write briefly about Biodegradable polymers.
9. Explain about conducting polymers.
10. Write note on the following:
    (*a*) Solar cells
    (*b*) Solar panels
    (*c*) Solar heaters
    (*d*) Solar reflectors
    (*e*) Greenhouse concept.
11. Define green chemistry. Explain the need and principles of green chemistry.
12. Explain the following methods for green synthesis:
    (*a*) Aqueous phase method
    (*b*) Super critical fluid extraction method
    (*c*) Phase transfer catalyst (PTC)
    (*d*) Biocatalyst for green synthesis
    (*e*) Microwave induced method
    (*f*) Ultrasound method
    (*g*) Ionic solvent method.
13. What is cement? How the portland cement is manufacture by rotary kiln?
14. Write a note on setting and hardening of cement.
15. Write about the following:
    (*a*) Heat of hydration
    (*b*) Decay of cement.